THE INVISIBLE CENTURY

ALSO BY RICHARD PANEK

Seeing and Believing: How the Telescope Opened Our Eyes and Minds to the Heavens

Waterloo Diamonds

THE INVISIBLE CENTURY

EINSTEIN, FREUD, AND THE SEARCH FOR HIDDEN UNIVERSES

RICHARD PANEK

VIKING

VIKING
Published by the Penguin Group
Penguin Group (USA) Inc., 375 Hudson Street,
New York, New York 10014, U.S.A.
Penguin Books Ltd, 80 Strand, London WC2R 0RL, England
Penguin Books Australia Ltd, 250 Camberwell Road, Camberwell,
Victoria 3124, Australia
Penguin Books Canada Ltd, 10 Alcorn Avenue, Toronto, Ontario,
Canada M4V 3B2
Penguin Books India (P) Ltd, 11 Community Centre, Panchsheel Park,
New Delhi–110 017, India
Penguin Group (NZ), Cnr Airborne and Rosedale Roads, Albany,
Auckland 1310, New Zealand
Penguin Books (South Africa) (Pty) Ltd, 24 Sturdee Avenue, Rosebank,
Johannesburg 2196, South Africa

Penguin Books Ltd, Registered Offices: 80 Strand, London WC2R 0RL, England

First published in 2004 by Viking Penguin,
a member of Penguin Group (USA) Inc.

1 3 5 7 9 10 8 6 4 2

LIBRARY OF CONGRESS CATALOGING-IN-PUBLICATION DATA
Panek, Richard.
The invisible century : Einstein, Freud and the search for hidden universes /
by Richard Panek.
p. cm.
ISBN 0-670-03074-0
1. Science—Methodology. 2. Science—Philosophy. 3. Einstein, Albert,
1879–1955. 4. Freud, Sigmund, 1856–1939. I. Title.

Q175.P3325 2004
501—dc22

2003069463

This book is printed on acid-free paper. ♾

Printed in the United States of America

Once again, for Meg Wolitzer, with love

ACKNOWLEDGMENTS

The author would like to thank Dawn Drzal for her help in formulating the proposal for this book, Barbara Grossman for her instant support, Christopher Potter for his early and ongoing enthusiasm, Wendy Wolf for her expert editorial guidance and patience and faith (and patience), Beena Kamlani for her hidden universe of editorial wisdom, Henry Dunow for his agent's counsel and perseverance, Gino Segrè for his advice on Einstein, Robert Galatzer-Levy for his advice on Freud, the Corporation of Yaddo for its generosity with time and space, Joel Whitebook for his hours of inspiration, and finally, for indulging their father yet one more time, Gabriel and Charlie.

CONTENTS

Prologue | 1

I. MIND OVER MATTER

One: More Things in Heaven | 9
Two: More Things on Earth | 35
Three: Going to Extremes | 55

II. MATTER OVER MIND

Four: A Leap of Faith | 81
Five: The Descent of a Man | 115

III. THE TREMBLING OF THE DEWDROP

Six: A Discourse Concerning Two New Sciences | 153

Notes | 209
Bibliography | 233
Index | 249

Q: "Is the invisible visible?"
A: "Not to the eye."

— from an 1896 interview with
Wilhelm Conrad Röntgen,
the discoverer of the X-ray

PROLOGUE

They met only once. During the New Year's holiday season of 1927, Albert Einstein called on Sigmund Freud, who was staying at the home of one of his sons in Berlin. Einstein, at forty-seven, was the foremost living symbol of the physical sciences, while Freud, at seventy, was his equal in the social sciences, but the evening was hardly a meeting of the minds. When a friend wrote Einstein just a few months later suggesting that he allow himself to undergo psychoanalysis, Einstein answered, "I regret that I cannot accede to your request, because I should like very much to remain in the darkness of not having been analyzed." Or, as Freud wrote to a friend regarding Einstein immediately after their meeting in Berlin, "He understands as much about psychology as I do about physics, so we had a very pleasant talk."

Freud and Einstein shared a native language, German, but their respective professional vocabularies had long since diverged, to the point that they now seemed virtually irreconcilable. Even so, Freud and Einstein had more in common than they might have imagined. Many years earlier, at the beginning of their respective scientific investigations, they both had reached what would prove to be the same pivotal juncture. Each had been exploring one of the foremost problems in his field. Each had found himself confronting an obstacle that had defeated everyone else exploring the problem. In both their cases, this obstacle was the same: a lack of more evidence. Yet rather than retreat from this absence and look elsewhere or concede defeat and stop looking, Einstein and Freud had kept looking anyway.

Looking, after all, was what scientists did. It was what defined the scientific method. It was what had precipitated the Scientific Revolution, some three centuries earlier. In 1610, Galileo Galilei reported that upon looking through a new instrument into the celestial realm he saw forty stars in the Pleiades cluster where previously everyone else had seen only six, five hundred new stars in the constellation of Orion, "a congeries of innumerable stars" in another stretch of the night sky, and then, around Jupiter, moons. Beginning in 1674, Antonius von Leeuwenhoek reported that upon looking at terrestrial objects through another new instrument he saw "upwards of one million living creatures" in a drop of water, "animals" numbering more than "there were human beings in the united Netherlands" in the white matter on his gums, and then, in the plaque from the mouth of an old man who'd never cleaned his teeth, "an unbelievably great number of living animalcules, a-swimming more nimbly than any I had ever seen up to this time."

Such discoveries were not without precedent. They came, in fact, at the end of the Age of Discovery. If an explorer of the seas could discover a New World, then why should an explorer of the heavens not discover new worlds? And if those same sea voyages proved that the Earth could house innumerable creatures previously unknown, then why not earth itself or water or flesh?

What *was* without precedent in the discoveries of Galileo and Leeuwenhoek, however, was the means by which they reached them. Between 1595 and 1609, spectacle makers in the Netherlands had fit combinations of lenses together in two new instruments that performed similar, though distinct, optical tricks. The combination of lenses in one instrument made distant objects appear nearer, the combination in the other made small objects appear larger; and for the first time in history investigators of nature had at their disposal tools that served as an extension of one of the five human senses. As much as the discoveries themselves, what revolutionized science over the course of the seventeenth century was a new means of discovery and what it signified: There is more to the universe than meets the naked eye.

Who knew? After all, these instruments might easily have revealed nothing beyond what we already knew to be there, and what we already knew to be there might easily have been all there was to know. The naked eye alone didn't *have* to be inadequate as a means of investigating nature; the invention of these instruments didn't *have* to open two new frontiers. But it was; and they did.

For thousands of years, the number of objects in the heavens had been fixed at six thousand or so. Now, there were . . . more. Since the Creation, or at least since the Flood, the number of kinds of creatures on Earth, however incalculable as a practical matter, had nonetheless been fixed. Now, there were . . . more. "There are more things in heaven and earth,

Horatio, than are dreamt of in your philosophy": When Shakespeare wrote these words in 1598 or 1599, at the very cusp of the turn of the seventeenth century, he was referring to the understandable assumption among practitioners of what would soon become the old philosophy that much of what was as yet unknown must remain unknown forever, and for the next three hundred years the practitioners of what they themselves came to call the New Philosophy frequently cited it as the last time in history that someone could have written so confidently about civilization's continuing ignorance of, and estrangement from, the universe.

Because now all you had to do was look. Through the telescope you could see farther than with the naked eye alone and, by seeing farther, discover new worlds without. Through the microscope you could see deeper than with the naked eye alone and, by seeing deeper, discover new worlds within. By seeing more than meets the naked eye and then seeing yet more, you could discover more.

How much more? It was a logical question for natural philosophers to ask themselves, and the search for an answer that ensued over the next three centuries was nothing if not logical: a systematic pursuit of the truths of nature to the outermost and innermost realms of the universe, until by the turn of the twentieth century the search had reached the very limits of human perception even with the aid of optical instruments, and investigators of nature had begun to wonder: What now? What if there was no more *more*?

Specifically: Was the great scientific program that had begun three centuries earlier coming to a close? Or would an increasingly fractional examination of the existing evidence continue to reward investigators with further truths?

Some researchers, however, unexpectedly found themselves confronting a third option. Pushing the twin frontiers of scientific research—the inner universe and the outer—they had arrived at an impasse. Then, they'd spanned it. They'd kept looking until they discovered an entirely new kind of scientific evidence: evidence that no manner of mere looking was going to reveal; evidence that lay beyond the realm of the visible; evidence that was, to all appearances, invisible.

The invisible had always been part of humanity's interactions with nature. Attempting to explain otherwise inexplicable phenomena, the ancients had invented spirits, forms and gods. In the Western world during the medieval era, those various causes of mysterious effects had coalesced into the idea of one God. Even after the inception of the modern era and the inauguration of the scientific method, investigators working at the two extremes of the universe had resorted to two new forms of the invisible. When Isaac Newton reached the limits of his understanding of the outer universe, he had invoked the concept of gravity. When René Descartes reached the limits of his understanding of the inner universe, he had invoked the concept of consciousness.

But by the turn of the twentieth century the kind of invisibility that certain investigators were beginning to invoke was new. These were scientists for whom any appeal to the supernatural, superstitious, or metaphysical would have been anathema. But now, here it was: evidence that was invisible yet scientifically incontrovertible, to their minds, anyway.

Although Einstein and Freud didn't initiate this second scientific revolution all by themselves, they did come to represent it and in large measure embody it. This is the story of how their respective investigations reached unprecedented realms, relativity and the unconscious; how their further pursuits led to the somewhat inadvertent creation of two new sciences, cosmology and psychoanalysis; and how in Einstein's case, a new way of doing science has become the dominant methodology throughout the sciences, while in Freud's case, an alternative way of doing science has become the dominant exception, the key to the very question of what qualifies an intellectual endeavor as a science. This is also the story of what cosmology and psychoanalysis have allowed us to explore: universes, without and within, as vast in comparison to the ones they replaced as those had been to the ones *they* replaced.

And in that regard Einstein and Freud's is a story, just as Galileo and Leeuwenhoek's was, of a revolution in thought. The difference between our vision of the universe and its nineteenth-century counterpart has turned out to be *not* a question of what had distinguished each previous era from the preceding one for nearly three hundred years: of seeing farther or deeper, of seeing *more*—of perspective, of how much we see. Instead, it is a question of seeing itself—of perception, of *how* we see. It is also, then, a question of *thinking* about seeing—of conception, of how we think about how we see. As much as any discovery, this is what has changed the way we try to make sense of our existence in the twenty-first century—the way we struggle to investigate our circumstances as sentient creatures in a particular setting: Who are these creatures? What is this setting? It is a new means of discovery—the significance of which, a hundred years later, we are still only beginning to comprehend: that there is more to the universe than we would ever find, if all we ever did was look.

I

MIND OVER MATTER

ONE

MORE THINGS IN HEAVEN

Look.

And so the boy looked. His father had something to show him. It was small and round like a miniature clock, the boy saw, but instead of two hands pointing outward from the center of the face it had one iron needle. As the boy continued to look, his father rotated the object. He turned it first one way, then the other, and as he did so the most amazing thing happened. No matter how the boy's father moved the object, the needle continued pointing in the same direction—not the same direction relative to the rest of the device, as the boy might have expected, but the same direction relative to . . . something else. Something out there, outside the device, that the boy couldn't see. The needle was shaking now. It trembled with the effort. Some six decades later, when Albert Einstein recalled this scene, he couldn't remember whether he had been four or five at the time, but the lesson he'd learned that day he could still summon and summarize crisply: "Something deeply hidden had to be behind things."

Some *things* deeply hidden, actually. As the boy grew older, he learned what a few of those deeply hidden somethings were: magnetism, the subject of his father's demonstration on that memorable day; electricity; and the relationship between the two. He learned that the existence of a relationship between magnetism and electricity was still so recent a discovery that nobody yet understood how it worked, and then he learned that within his lifetime physicists had demonstrated that this relationship manifested itself to our eyes as light. And he learned that even though nobody yet understood how light worked, what everybody did know was that it traveled along the biggest deeply hidden something of them all, one that had so far eluded the greatest minds of the age but one that was now, as a prominent physicist of the era proclaimed, "all but in our grasp."

That something was the ether. Einstein himself sought it, in a paper he wrote in 1895, *"Über die Untersuchung des Ätherzustandes im magnetischen Felde"* ("On the Investigation of the State of the Ether in a Magnetic Field"). This contribution to the literature, however, wasn't so much original scholarship as a five-finger exercise that wound up pretty much reiterating current thinking, since Einstein was only sixteen at the time and, as he cautioned prospective readers (such as the doting uncle to whom he sent the paper), "I was completely lacking in materials that would have enabled me to delve into the subject more deeply than by merely meditating on it."

Still, it was a start. Over the following decade, Einstein would graduate from self-consciously precocious adolescent, speculating beyond his abilities, to willfully arrogant student at the Swiss Polytechnic in Zurich, to humble (if not quite humbled) clerk at the Swiss Office for Intellectual Property in Bern, where he wound up in part because his professors had refused to write letters of recommendation for someone so dismissive of their authority. As Einstein reported in a letter in May 1901, "From what I have been told, I am not in the good books of any of my former teachers." Yet even as a patent clerk, Einstein continued to seek the ether, for the same reason that physicists everywhere were seeking the ether. When electromagnetic waves of light departed from a star that was *there* and hadn't yet arrived *here,* they had to be traveling along *something.* So: What was it? Find that something, as physicists understood, and maybe electricity and magnetism and the relationship between the two wouldn't seem so deeply hidden after all.

Among the seekers of the ether, one was without equal: the Scottish physicist William Thomson, eventually Baron Kelvin of Largs. As one of the most prominent and illustrious physicists of the century, Lord Kelvin had made the pursuit of the ether the primary focus of his scientific investigations for literally the entire length of his long career. He'd first thought he found it on November 28, 1846, during his initial term as a professor of natural history at the University of Glasgow. He was mistaken. As he wrote a friend in 1896, on the occasion of the golden jubilee of his service to the university, a three-day celebration that attracted two thousand representatives of scientific societies and academies of higher learning from around the world, "I have not had a moment's peace or happiness in respect to electromagnetic theory since Nov. 28, 1846."

Part of the problem with the ether was how to picture it. "I never satisfy myself unless I can make a mechanical model of a thing," Kelvin once told a group of students. "If I can make a mechanical model, I can understand it." In one such demonstration that was a perennial favorite of his students, he would draw geometrical shapes on a piece of india rubber, stretch the rubber across the ten-inch mouth of a long brass funnel, and, having hung the funnel upside down over a tub, direct water from a supply pipe into the thin tube at the top. As the water collected in the mouth of the funnel, the india rubber bulged, and it drooped, and it gradually assumed the shape of a globule. Soon the blob had expanded to a width nearly double the diameter of the mouth from which it appeared to be emerging, and just when the rubber seemed unable to stretch any thinner, it did anyway. All the while Kelvin continued to lecture, calmly commenting on the subject of surface tension as well as on the transformations the simple Euclidean shapes on the rubber were now undergoing. Then, at precisely the moment Kelvin calculated that neither india rubber nor the ten benches of physics students could endure any greater tension, he would raise his pointer, poke the gelatinous mass hanging before him, and, turning to the class, announce, "The trembling of the dewdrop, gentlemen!"

The trembling of the dewdrop, the angling of the gas molecule, the orbiting of a planet: the least matter in the universe to the greatest, and all operating according to the same unifying laws. Here was the whole of modern science, in one easy lesson. More than two hundred years earlier, René Descartes had expressed the philosophical hope that a full description of the material universe would require nothing but matter and motion, and several decades after that Isaac Newton had expressed the physical principles that described the motion of matter. The rest, in a way, had been a process of simply filling in the blanks—plugging measurements of matter into equations for motions, and watching the universe tumble out piecemeal yet unmistakably all of a single great mechanistic piece. The lecture hall where for half a century Kelvin demonstrated his models was a monument of sorts to this vision: the triple-spiral spring vibrator he'd hung from one end of the blackboard; the thirty-foot pendulum, consisting of a steel wire and a twelve-pound cannonball, that he'd suspended from the apex of the dome roof; two clocks, those universal symbols of the workings of the universe. Matter and motion, motion and matter, one acting upon the other; causes leading inexorably to effects that, by dint of more and more rigorous and precise examination, were equally predictable and verifiable to whatever degree of accuracy anyone might care to name: Here was a cosmos complete, almost.

The exception was the ether. When numerous experiments in the early nineteenth century began showing that light travels in waves, physicists naturally tried to describe a substance capable of carrying those waves. The consensus: an absolutely incompressible, or elastic, solid. For Cambridge physicist George Gabriel Stokes, that description suggested a combination of glue and water that would act as a conduit for rapid vibrations of waves and also allow the passage of slowly moving bodies. For British physicist Charles Wheatstone, it meant white beads, which he used in his Wheatstone wave machine of the early 1840s—a visual aid that vividly demonstrated how ether particles might move at right angles to a wave coursing through their midst and an inspiration for numerous similar teaching aids of the era.

And for Kelvin, "the nearest analogy I can give you," as he once said during a lecture, "is this jelly which you see." On other occasions, he might begin his demonstration with

Scotch shoemakers' wax. If he shaped the wax into a tuning fork or bell and struck it, a sound emanated. Then he would take that same sound-wave-conveying wax and suspend it in a glass jar filled with water. If he first placed corks under the substance, then laid bullets across the top of it, in time the positions of the objects would reverse themselves. The bullets would sink through the wax to the bottom while the corks would pop out the top. "The application of this to the luminiferous ether is immediate," he concluded: a substance rigid enough to conduct waves traveling at fixed speeds in straight lines from one end of the universe to the other, if need be, yet porous enough not to block the passage of bullets, corks, or even—by the same application of scale that rendered minuscule dewdrops and giant rubber globules analogous—planets.

Not to block—but surely to impede? Surely at least to *slow* the passage of a planet? An elastic solid occupying all of space would *have* to present a degree of resistance to a (in the parlance of the day) "ponderable body" such as Earth. But to what degree precisely? In an effort to determine the exact extent of the luminiferous (or light-bearing) ether's drag on Earth, the American physicist Albert A. Michelson devised an experiment that he first conducted in Berlin in 1881. His idea was to send two beams of light along paths at 90-degree angles to each other. Presumably the beam following one path would be fighting *against* the current as Earth plowed through the ether, while the beam on the other path would be swimming *with* the current. Michelson designed an ingenious instrument, which he called an interferometer, that he hoped would allow him to make measurements that, through a series of calculations, would determine the velocity of the Earth through the ether.

The Berlin reading, however, suffered from the vibrations of the horse cabs passing outside the Physical Institute. So he moved his apparatus to the relative isolation of the Astrophysical Observatory in Potsdam, where he repeated the experiment. The reading, to his surprise, indicated nothing.

Which was impossible. An interaction between a massive planet and even the most elastic of solids surely couldn't pass undetected or remain undetectable. "One thing we are sure of," Kelvin told an audience in Philadelphia three years later, while on his way to lecture at Johns Hopkins University in Baltimore, "and that is the reality and substantiality of the luminiferous ether." And if experiments of unprecedented refinement and sophistication failed to detect it, there was only one reasonable alternative course of action. As Kelvin wrote in his preface to the published volume of those Baltimore Lectures, "It is to be hoped that farther experiments will be made."

They were. In 1887 Michelson tried again, this time with the help of the chemist Edward W. Morley. Together they constructed an interferometer far more elaborate and sensitive than the ones Michelson had used in Germany, secured it in an essentially tremor-free basement at the Case School of Applied Science in Cleveland, and set it floating on a bed of mercury for, literally, good measure. Michelson had in mind a specific number for the wavelength displacement he expected the ether would produce, and he further decided that a reading 10 percent of that number would conclusively indicate a null result. What he got was a reading of 5 percent of the displacement he thought the ether might produce—a blip attributable to observational error, if anything. Michelson found himself forced to reach the same conclusion he'd previously reported:

"that the luminiferous ether is entirely unaffected by the motion of the matter which it permeates."

"I cannot see any flaw," said Kelvin of this experiment, in a lecture he delivered in the summer of 1900. "But a possibility of escaping from the conclusion which it seemed to prove may be found in a brilliant suggestion made independently by FitzGerald, and by Lorentz of Leiden." Kelvin was referring to the physicists George Francis FitzGerald of Dublin, who had submitted a brief conjecture regarding the ether to the American journal *Science* in 1889, and Hendrik Antoon Lorentz, who in an 1892 paper and then in an 1895 book-length treatise had elaborated an entire argument along nearly identical lines: The ether compresses the molecules of the interferometer—as well as those of the Earth, for that matter—to the exact degree necessary to render a null result. In which case, the two beams of light in Cleveland actually did travel at two separate speeds, as the measurements of their multiple-mirror-deflected journeys would have shown, if only the machinery hadn't contracted just enough to make up the difference. "Thus," Lorentz concluded, "one would have to imagine that the motion of a solid body (such as a brass rod or the stone disc employed in the later experiments) through the resting ether exerts upon the dimensions of that body an influence which varies according to the orientation of the body with respect to the direction of motion."

"An explanation was necessary, and was forthcoming; they always are," the French mathematician and philosopher Henri Poincaré wrote of Lorentz in 1902 in his *Science and Hypothesis;* "hypotheses are what we lack the least." Lorentz himself conceded as much. Two years later he proposed a mathematical basis for his argument while virtually sighing at the futility of the whole enterprise: "Surely this course of inventing special hypotheses for each new experimental result is somewhat artificial."

Like other physicists at the time, Einstein thought about ways to describe the ether, as in the precocious paper he had sent to his uncle in 1895. Also like other physicists, Einstein thought about ways to detect the ether. During his second year at college, 1897–98, he proposed an experiment: "I predicted that if light from a source is reflected by a mirror," he later recalled, "it should have different energies depending on whether it is propagated parallel or antiparallel to the direction of motion of the Earth." In other words: the Michelson-Morley experiment, more or less—though news of that effort, a decade earlier, had reached Einstein only indirectly if at all, and then only as a passing reference in a paper he read. In any case, the particular professor he'd approached with this proposal treated it in "a stepmotherly fashion," as Einstein reported bitterly in a letter. Then, during a brief but busy job-hunting period in 1901, after he'd left school but hadn't yet secured a position at the patent office, Einstein proposed to a more receptive professor at the University of Zurich, "a very much simpler method of investigating the relative motion of matter against the luminiferous ether." On this occasion it was Einstein who didn't deliver. As he wrote to a friend, "If only relentless fate would give me the necessary time and peace!"

Like a few other physicists at the time, Einstein was even beginning to wonder just what purpose the ether served. What purpose it was *supposed* to serve was clear enough. Physicists had inferred the ether's existence in order to make the discovery of light waves conform to the laws of mechanics. If the universe operated only through matter moving immediately adjacent matter in an endless succession of cause-and-effect ricochet shots—like balls on a billiard table, in the popular analogy of the day—then the ether would serve as the necessary matter facilitating the motion of waves of light across the vast and otherwise empty reaches of space. But to say that the ether is the substance along which electromagnetic waves must be moving because electromagnetic waves must be moving along *something* was as unsatisfactory a definition as it was circular. As Einstein concluded during this period in a letter to the fellow physics student who later became his first wife, Mileva Maric, "The introduction of the term 'ether' into the theories of electricity led to the notion of a medium of whose motion one can speak without being able, I believe, to associate a physical meaning with this statement."

The problem of the ether was starting to seem more than a little familiar. It was, in a way, the same problem that had been haunting physics since the inception of the modern era three centuries earlier: space. To be precise, it was absolute space—a frame of reference against which, in theory, you could measure the motion of any matter in the universe.

For most of human history, such a concept would have been more or less meaningless, or at least superfluous. As long as Earth was standing still at the center of the universe, the center of the Earth was the rightful place toward which terrestrial objects must fall. After all, as Aristotle pointed out in establishing a comprehensive physics, that's precisely what terrestrial objects did. An Earth in motion, however, presented another set of circumstances altogether, one that—as Galileo appreciated—required a whole other set of explanations.

Nicolaus Copernicus wasn't the first to suggest that the Earth goes around the sun, not vice versa, but the mathematics in his 1543 treatise *De revolutionibus orbium coelestium* (*On the Revolutions of Celstial Orbs*) had the advantage of being comprehensive and even useful—for instance, in instituting the calendar reform of 1582. Still, for many natural philosophers its heliocentric thesis remained difficult, or at least politically unwise, to believe. Galileo, however, not only found it easy to believe but, in time, learned it had to be true because he had seen the evidence for himself, through a new instrument that made distant objects appear near. His evidence was not the mountains on the moon that he first observed in the autumn of 1609, though they did challenge one ancient belief, the physical perfection of heavenly bodies; nor the sight of far more stars than were visible with the naked eye, though they did hint that the two-dimensional celestial vault of old might possess a third dimension; not even his January 1610 discovery around Jupiter of "four wandering stars, known or observed by no one before us," because all they proved was that Earth wasn't unique as a host of moons or, therefore, as a center of rotation. Instead, what finally decided the matter for Galileo was the phases of Venus. From October to December 1610, Galileo mounted a nightly vigil to observe Venus as it mutated from "a round shape, and very small," to "a semicircle" and much larger, to "sickle-shaped" and very large—exactly the set of appearances the planet would manifest if it were circling around, from behind the sun to in front of the sun, while also drawing nearer to Earth.

Galileo's discovery of the phases of Venus didn't definitively prove the existence of a sun-centered universe. It didn't even necessarily disprove an Earth-centered universe. After all, just because Venus happens to revolve around the sun doesn't mean that the sun itself can't still revolve around Earth. But such a contortionistic interpretation of the cosmos—a Venus-encircled sun in turn circling Earth—had nothing to recommend it other than an undying allegiance to Earth's central position in it. And so "Venus revolves around the Sun," Galileo finally declared with virtual certainty, in a letter he wrote in January 1612 and published the following year, "just as do all the other planets"—a category of celestial object that, he could now state with a confidence verging on nonchalance, included the heretofore terrestrial-by-definition Earth.

An Earth spinning and speeding through space, however, required not only a rethinking of religious beliefs. It also required new interpretations of old physical data—a new physics. Galileo himself got to work on one, and in 1632 he published it: *Dialogue Concerning the Two Chief World Systems.* In arguing on behalf of a Copernican view of the universe, Galileo knew he was going to have to explain certain phenomena that in the Aristotelian view of the universe needed no further explanation. Actually, he was going to have to explain the *absence* of certain phenomena: If Earth were turning and if this turning Earth were orbiting the sun, as Copernicus contended, then wouldn't birds be rent asunder, cannonballs be sent off course, and even simple stones, dropped from a modest height, be flung far from their points of departure, all according to the several motions of the planet?

No, Galileo said. And here's why. He asked you, his reader, to imagine yourself on a dock, observing a ship anchored in a harbor. If someone at the top of the ship's mast were to drop a stone, where would it land? Simple: at the base of the mast. Now imagine instead that the ship is moving in the water at a steady rate across your field of vision as you observe from the dock. If the person at the top of the ship's mast were to drop another stone, where would it land now? At the base of the mast? Or some small distance back, behind the mast—a measurement corresponding to the distance on the water that the ship would have covered in the time between the release of the stone at the top of the mast and its arrival on the deck of the ship?

The intuitive, Aristotelian answer: some small distance back. The correct—and, Galileo argued, Copernican—answer: the base of the mast, because the movement of the ship and the movement of the stone together constitute *a single motion.* From the point of view of the person at the top of the mast, the motion of the stone *alone* might indeed seem a perpendicular drop—the kind that Aristotle argued a stone would make in seeking to return to its natural state in the universe. Fair enough. That's what it would have to seem to someone standing on the steadily moving ship whose only knowledge of the motion of the Earth was that it stood still. That person would feel neither the motion of the Earth nor the motion of the ship and so would take into account only the motion of the stone. But for you, observing from the dock, the stone would be moving *and* the ship would be moving, and *together* those movements would make up a single system in motion. To you, the motion of the stone falling toward the ship would seem not a perpendicular drop—not at all an Aristotelian return to its natural state—but an angle. If you could trace the trajectory of the stone from the dock, it would just be geometry.

And vice versa. If, instead, you the observer standing on the dock were the one dropping a stone, then to you the motion of that stone relative to the Earth would appear perpendicular, because all you would be taking into account was the motion of the stone *alone.* That's all Aristotle did—take into account only the motion of the stone. But from the point of view of the person at the top of the mast on the ship in the harbor, looking at you on the dock and taking into account the motion of the stone and the apparent motion of the dock *together,* the trajectory of the falling stone would describe an angle.

And there it is: a principle of relativity. Neither observer would have the right to claim to be absolutely at rest. The onboard observer would have as much right to claim that the ship was leaving the dock as that the dock was leaving the ship. Rather than standing still at the center of the cosmos, our position in the new physics was just the opposite: never at rest. After Galileo, everything in the universe was in motion relative to something else—ships to docks, moons to planets, planets to sun, sun (as astronomers would come to discover by the end of the eighteenth century) to the so-called fixed stars, those so-called fixed stars (as astronomers would come to discover by the middle of the nineteenth century) to one another, and, conceivably, our entire vast system of stars (as astronomers were trying to determine at the turn of the twentieth century) to other vast systems of stars.

Unless you counted the ether. For this reason alone, the ether was—as Einstein had first recognized as a teenager—at least somewhat objectionable. Not long after he'd written the ether paper that he'd sent to his uncle, Einstein found himself wandering the grounds at his school in Aarau, Switzerland, wondering what the presence of an absolute space would do to Galileo's idea of relativity. If you were on board Galileo's ship but belowdecks, in an enclosed compartment, you shouldn't be able to detect whether you were moving or standing still, relative to the dock or anything else in the universe that wasn't moving along with you. But if the ship were traveling at the speed of light through the ether, that's just what you *would* be able to detect. You'd know *you* were the one traveling at the speed of light—rather than someone on the dock, for instance—because you'd see the light around you standing still.

By the early years of the twentieth century, Einstein had done only what other physicists of his era had done. He'd thought about ways to define the ether through mathematics. He'd thought about ways to detect the ether through experiments. He'd even begun to think about whether physics really needed an ether. But then, one night in May 1905, Einstein did what no other physicist of his era had done. He thought of a new way of thinking about the problem.

Einstein had been spending the evening with a longtime friend both from his student years and at the patent office, Michele Besso, the two of them talking, as they often did in their off-hours, about physics. In the preceding three years, Einstein had moved to Bern, gotten married, and fathered two children (one illegitimate, whom he and Mileva gave up for adoption). Yet all the while he'd been applying himself to the most pressing issues of contemporary physics, often in the company of his patent-office sounding board, Besso. On this particular occasion, Einstein had approached Besso for the express purpose of doing "battle" with a problem that had been plaguing him on and off for the past decade. After a lively discussion, Einstein returned home, where, all at once, he understood what he and everyone else who had been studying the situation had been overlooking all along.

"Thank you!" he greeted Besso the following day. "I have completely solved the problem." The trouble with the current conception of the universe, he explained, wasn't absolute space—or at least wasn't *only* absolute space. It was absolute time.

"If, for example, I say that 'the train arrives here at 7 o'clock,' that means, more or less, 'the pointing of the small hand of my watch to seven and the arrival of the train are simultaneous events.'" This sentence comes early in *"Zur Elektrodynamik bewegter Körper"* ("On the Electrodynamics of Moving Bodies"), the paper that Einstein completed and mailed to the *Annalen der Physik* six weeks later. In its audacious simplicity, even borderline simplemindedness, this sentence is deceptive, for with this description of one of the most mundane of human observations—one that just about any eight-year-old can make—Einstein pinpointed precisely what everyone else who had been studying the problem had missed: "time" is not universal or absolute; it is not sometimes universal and sometimes local or relative; it is *only* local.

The key was the speed of light. The fact that the speed of light is not infinite, as Aristotle and Descartes and so many other investigators of nature over the millennia had supposed, had been common knowledge since the late seventeenth century. So had its approximate value. In 1676, the Danish astronomer Ole Rømer used the data from years of observations at the Paris Observatory to determine that the timing of the eclipses of Jupiter's innermost moon depended on where Jupiter was in its orbit relative to Earth. The eclipses came earlier when Earth was nearest Jupiter, later when Earth was farthest from Jupiter, suggesting that the eclipses didn't happen at the very same moment we saw them happen. That, in fact, *when* we saw them depended on *where* they happened, nearer or farther. "This can only mean that light takes time for transmission through space," Rømer concluded—140,000 miles per second, by the best estimates of the day.

But the combination of these two factors—that the speed of light is incomprehensibly fast; that the speed of light is inarguably finite—didn't begin to assume a literally astronomical dimension for another hundred years. Beginning in the 1770s, William Herschel (the same observer who proved that the sun is in motion relative to the fixed stars) began systematically exploring the so-called celestial vault—the ceiling of stars that astronomers had known since Galileo's time must have a third dimension but that they still couldn't help conceiving as anything except a flat surface. With every improvement in his telescopes, Herschel pushed his observations of stars to greater and greater depths in the sky or distances from Earth or—since the speed of light coming from the stars *is* finite, since it does *take time* to reach our eyes—farther and farther into the past. "I have looked further into space than ever human being did before me," Herschel marveled in 1813, in his old age. "I have observed stars of which the light, it can be proved, must take two million years to reach the earth."

Even that distance, however, would seem nearby if the speculations of some astronomers at the turn of the twentieth century turned out to be true. If certain smudges at the farthest reaches of the mightiest telescopes turned out to be systems of stars outside our own—other "island universes" altogether equal in size and magnitude to our own Milky Way—then when we looked at the starlight reaching us from them we might be seeing not Herschel's previously unfathomable two million years into the past but two hundred million years or even two thousand million years. And so they would go, these meditations on the meaning of light, ever and ever outward, further and further pastward, if not necessarily *ad infinitum,* then at least, quite possibly, *ad absurdum.*

Now Einstein reversed that trajectory. Instead of considering the implications of looking farther and farther across the universe and thereby deeper and deeper into the past, he thought about the meaning of looking nearer and nearer—or, by the same reasoning, closer and closer to the present. Look near enough, he realized, and you'll be seeing very close indeed to the present. But only one place can you claim to *be* the present—and then only *your* present.

It was this insight that allowed Einstein to endow the idea of time with an unprecedented immediacy, in both the positional and the temporal senses of the word: here and now: the arrival of a train *and* the hands of a watch. Because the train and the hands of the watch occupy the same location, they also occupy the same time. For an observer standing immediately adjacent to the train, that time is, by definition, the present: seven o'clock. But someone in a different location observing the arrival of that same train—that is, someone at some distance away receiving the image of the train, which has traveled by means of electromagnetic waves from the surface of the locomotive to the eyes of this second observer at the speed of light, an almost unimaginably high *yet nonetheless finite* velocity—wouldn't be able to consider the arrival of the train simultaneous with its arrival for the first observer. If light did propagate instantaneously—if the speed of light were in fact infinite—then the two observers *would* be seeing the arrival of the train simultaneously. And indeed, it might very well *seem* to them as if they were, especially if (using the modern value for the speed of light as 186,282 miles, or 299,792 kilometers, per second) the other observer happens to be standing on a street corner that's about two-millionths of a light-second (the distance that light travels in two-millionths of a second, or slightly less than 2000 feet) away rather than, less ambiguously, near a star that's two million light-years (or slightly less than 12 quintillion miles) away. And yes, if you were the observer on the street corner in the same town, gazing down a hill at a slowing locomotive pulling into the station, the arrival of the train for all practical purposes *might as well be* happening at the same moment as its arrival for the observer on the platform.

But what it is, is in your past.

Einstein was not in fact alone in recognizing the role that the velocity of light plays in the conception of time. Other physicists and philosophers had begun to note a paradox at the heart of the concept of simultaneity—that for two observers, the difference in distances has to translate into a difference in time. But where Einstein diverged from even the most radical of his contemporaries was in accepting as potentially decisive what the velocity of light is.

It was there in the math. In 1821, the British physicist Michael Faraday had decided to investigate reports from the Continent concerning electricity and magnetism by placing a magnet on a table in his basement workshop and sending an electrical impluse through a wire dangling over it. The wire began twirling, as if the electricity were sparking downward and the magnetism were influencing upward. This was, in effect, the first dynamo, the invention that would drive the industrial revolution for the rest of the century, and the product that Einstein's own father and uncle would manufacture as the family business. But not until the 1860s did the Scottish physicist James Clerk Maxwell manage to capture Faraday's accomplishment in mathematical form, a series of equations with an unforeseen implication. Electromagnetic waves travel at the same speed as light (and therefore, Maxwell predicted, *are* light): 186,282 miles, or 299,792 kilometers, per second in a vacuum. Meaning . . . what? That it would be *more* than

186,282 miles per second if you were moving away from the source of light, or *less* than 186,282 miles per second if you were moving toward the source? Yes—according to Newton's mechanics. Yet *it never seemed to vary.*

On a planet that was spinning; a spinning planet that was orbiting the sun; a spinning planet orbiting a sun that itself was moving in relation to other stars that were moving in relation to one another—in this setting that, as Copernicus and Galileo and Newton and Herschel and so many other astronomers and mathematicians and physicists and philosophers had so persuasively established, was never at rest and *therefore wouldn't be at rest in relation to a source of light outside itself,* always the answer to the question of what was the speed of light seemed to come up exactly the same. Just as Aristotelian philosophers considering the descent of an onboard stone would have overlooked the motion of the ship, so maybe several generations of Galilean physicists had been overlooking properties of electromagnetism. Maybe what you needed to consider was the motion of the stone, the motion of the ship, *and* the motion of the medium by which we perceive both, and together *those* three elements would constitute a single system in motion.

A few years earlier, his friend Besso had given Einstein a copy of the Austrian physicist Ernst Mach's *Die Mechanik in ihrer Entwicklung* (*The Science of Mechanics*). This work, Einstein later recalled, "exercised a profound influence upon me" because it questioned "mechanics as the final basis of all physical thinking." The issue for Mach wasn't whether mechanics had worked well over the past two centuries in describing the motions of matter; clearly, it had. The issue wasn't even whether mechanics could answer all questions about the physical universe, as the Kelvins of the world were constantly trying to prove. Rather, the issue for Mach—the root of his objection to Newtonian mechanics—was that it raised some questions it couldn't answer.

For instance, absolute space, the existence of which is necessary to measure absolute motions: On close reading, Newton's definition of it turned out to be every bit as circular as the reigning definition of the ether. "Absolute motion," Newton had written, "is the translation of a body from one absolute place into another." And what is place? "Place is a part of space which a body takes up, and is according to the space, either absolute or relative." So what, then, is absolute space? "Absolute space, in its own nature, without relation to anything external, remains always similar and immovable." Newton anticipated some criticism: "It is indeed a matter of great difficulty to discover, and effectually to distinguish, the true motions of particular bodies from the apparent; because the parts of that immovable space, in which those motions are performed, do by no means come under the observation of our senses. Yet the thing is not altogether desperate," he reassured the reader, "for we have some arguments to guide us, partly from the apparent motions, which are the differences of the true motions; partly from the forces, which are the causes and effects of the true motions." And what are these true, or absolute, motions? See above.

"We join with the eminent physicist Thomson [later Lord Kelvin] in our reverence and admiration of Newton," Mach wrote in 1883. "But we can only comprehend with difficulty his opinion that the Newtonian doctrines still remain the best and most philosophical foundation that can be given." Not that Mach was proposing an alternative to Newtonian mechanics; not that he was even suggesting physics was in need of an alternative. Rather, he was trying to remind his fellow physicists that just because mechanics had come "historically first" in modern science didn't mean that it had to be historically *final*. This was the argument that "shook" Einstein's "dogmatic faith" in mechanics alone as the basis of the physical world, and now, in May 1905, this was the argument that led Einstein to wonder whether mechanics and electromagnetism *together* could accommodate a principle of relativity—whether a synthesis of those two systems might in fact be historically *next*.

He tried it. First Einstein proposed that just as Newton's mechanics don't allow observers either on a dock or on a ship to consider themselves to be the ones absolutely at rest, neither should electrodynamics and optics. "We shall raise this conjecture (whose content will hereafter be called 'the principle of relativity') to the status of a postulate," he wrote in the second paragraph of his paper. Then he accepted the constancy of the speed of light in empty space as another given—a second postulate: that the speed of light in a vacuum is always the same "independent of the state of motion of the emitting body."

And that was all. It worked. Einstein now had two mutually reinforcing postulates, "only apparently irreconcilable": a principle of relativity, allowing us to conduct experiments involving light either on the ship or on the dock with equal validity; and a principle of constancy, allowing the ship (or the dock, for that matter) to approach the speed of light without any light onboard (or on the dock)—including electromagnetic waves bringing images from objects to our eyes—slowing to a stop and thereby revealing whether the ship (or dock) is the one "really" in motion. In which case, as Einstein promised his readers in that same second paragraph, the "introduction of a 'light ether' will prove to be superfluous, inasmuch as the view to be developed here will not require a 'space at absolute rest' endowed with special properties."

The rest was math—high-school-level algebra, at that. Suppose that you're standing on a dock watching our old Galilean ship, still anchored just offshore after all these centuries. And suppose that the ship is absolutely motionless in the water. And suppose that instead of dropping a stone, someone on board is dropping a light signal—sending a beam of light from the top of the mast to the deck. If you time this simple event and get an answer of, say, one second, then you know that the distance between the top of the mast and the deck must be the distance that light travels in one second, or 186,282 miles. (It's a big ship.)

The complications begin, just as they did in Galileo's day, once that ship lifts anchor and sets sail. Suppose that it's moving at a constant speed across your line of sight. If the person at the top of the mast sends a second light signal in the same manner as the first, what will you see from your vantage on the dock? The Aristotelian answer: a streak of light heading straight for the center of the Earth and therefore landing some distance behind the base of the mast—a distance corresponding to how far the ship has traveled along the water during the signal's journey. The Galilean answer: a streak of light heading straight for the base of the mast—which is the Einsteinian answer as well. From your point of view, the base of the mast will have moved out from under the top of the mast during the descent of the beam of light, just as it did during the descent of a stone. Which means the distance the light has traveled, from your point of view, has lengthened. It's not 186,282 miles. It's more.

How much more you can easily find out by measuring the time of its journey—and that's where the Einsteinian interpretation begins to depart from the Galilean. What is velocity? Nothing but distance divided by time, whether inches divided by—or, in the vernacular, *per*—day or kilometers per hour or miles per second. But if we accept Einstein's second postulate, then the velocity in question isn't just 186,282 miles per second. It's *always* 186,282 miles per second. It's constant—indeed, *a constant.* In the equation "velocity equals distance divided by time," this constant is over on one side of the equals sign, off by itself, humming along at its own imperturbable rate. On the opposite side of the equals sign are the parts of the equation that *can* vary, that are indeed *the variables*—distance and time, also known as miles and seconds. They can undergo as many permutations as you can imagine, as long as they continue to divide in such a way that the result is 186,282 miles per second, or the equivalent—372,564 miles per two seconds, or 558,846 miles per three seconds, or 1,862,820 per ten seconds, and so on. Change the distance, and you have to change the time.

You have to change the time.

For more than two centuries, though, you didn't. Now, on an evening in May 1905, you suddenly did, because on that evening Einstein, having talked the problem through with his friend Besso, realized that he needed to take into account something he'd never before adequately considered: the "inseparable connection between time and the signal velocity." Time was a variable, a measurement that passed at different rates according to where you were. To an observer on the dock, the second light signal would *had* to have lasted longer than one second. To an observer on the ship, however, the second light signal would have appeared to do what the first one had done, back when the ship was anchored in the water: travel straight down to the base of the mast, 186,282 miles away. For this observer, the distance wouldn't have differed from one signal to the next, so the time wouldn't have, either. The shipboard observer would be measuring one second while you on the dock would have been counting two seconds or three seconds or more, depending on the speed of the ship along the water. For this reason, you would have every right to say that clocks on board the ship were moving slowly. And there it is: a *new* principle of relativity.

Of course, such an effect wouldn't become noticeable unless the ship were moving at a significant fraction of the speed of light. At more modest speeds, the Galilean interpretation holds to a high degree of accuracy; as Einstein later wrote, "it supplies us with the actual motion of the heavenly bodies with a delicacy of detail little short of wonderful." Still, according to Einstein's math, as long as a ship is in motion *at all,* the distance the light travels on its angular path would have to be greater than the simple perpendicular drop you would see when the ship is at rest relative to you, and therefore the time to cover that distance would have to be greater, too. Through similar reasoning, Einstein also established that for an observer in a relative state of rest, back on the dock, the measurement of the length of a rod aboard a moving ship would have to shorten in the direction of motion, and to grow shorter the faster the ship is moving relative to the dock. And vice versa: Someone on the supposedly moving ship would have every right to consider *that* system to be the one at rest, and you and your so-called resting system to be the one whose dimensions would also appear to have shortened, and whose time would also appear to have slowed.

So which observer would be "right"? The observer on the ship, or you on the dock? The answer: Both—or, maybe more accurately, either, depending on who's doing the measuring. But how much time passed *really?* How long is the rod *really?* The answer: There is no "really"—no absolute space, no ether, against which to measure the motions of all matter in the universe. There is only the relative motions of the two systems.

"For the rest of my life I want to reflect on what light is," Einstein once said. If Einstein were correct, the universe wasn't quite a clockwork mechanism; it didn't function only according to the visible motions of matter. Instead it was electromagnetic, operating according to heretofore hidden principles. On a fundamental level, it was less a pocket watch than a compass.

Not that this new understanding of the universe was complete. Einstein knew that all he'd done was take into account the measurements of objects moving at uniform, or nonvarying, velocities relative to one another—a highly specialized situation. He hadn't yet taken into account the measurements of objects moving at nonuniform, or varying, velocities relative to one another—a far more representative sampling of the universe as we know it.

Still, it was a start. In a way, Einstein's light-centered universe was as physically distinct from the Galilean one he'd inherited as Galileo's sun-centered universe was from the Aristotelian one *he'd* inherited. But like Galileo, Einstein knew his had to be true—or truer than the one it was replacing, anyway—because he had seen the evidence for himself, if only in his mind's eye.

TWO

MORE THINGS ON EARTH

Listen.

And so the boy listened. His father had something to tell him. Hand in hand, they were going for a walk during which, in the manner they'd recently adopted, the father would attempt to impart to his son some lesson about life. On this occasion, the story concerned an incident that had happened to the father years earlier, on the streets of the city of Freiberg, the boy's birthplace. The father, then a young man, had been walking along minding his own business when a stranger came up to him and in one swift motion knocked his new fur hat off his head, called him a Jew, and told him to get off the pavement. The boy dutifully listened to his father describe this scene, and he had to wonder: So what did you do? His father answered quietly that he had simply stepped into the roadway and picked up his cap. The boy and his father then walked along in silence. The boy was considering this answer. He knew his father was trying to tell him something about how times had changed and how the treatment of Jews was better today. But that's not what the boy was thinking. Some three decades later, when Sigmund Freud recalled this scene, he couldn't remember whether he had been ten or twelve at the time, but the impression he'd taken away from that encounter he could still summon and summarize drily: "This struck me as unheroic conduct on the part of the big, strong man who was holding the little boy by the hand."

An impression, anyway. By the time Freud was committing this memory to paper, he was beginning to understand that any interpretation of the encounter on that long-ago day depended as much on what the boy had wanted to hear as on what the man had been trying to say, or even on what he, by now a father several times over, wished to believe about his father or himself, or about fathers and sons—depended, that is, on the vagaries of the human thought process. Not necessarily because of that conversation with his father (though maybe so; who knows?), it was the human thought process at the most basic level that Freud had grown up to explore: the pathways of nerves along which thoughts travel as they make their way through the brain. And so successful at tracing those paths were the neuroanatomists of Freud's generation that they seemed to have reached, as one contemporary historian of science declared, "the very threshold of mind."

That threshold was the neuron. Freud himself sought it, and in the early 1880s, as a young neuroanatomist fresh out of medical school, he even delivered a lecture before the Vienna Psychiatric Society about his own research into the structure of the nervous system. Although his subject that day was not specifically the ultimate point of connection between the fibers from two separate nerve cells within the human brain, he allowed himself a moment's speculation on what form such a juncture might take, though he immediately, and judiciously, added, "I know that the existing material is not sufficient for a decision on this important physiological problem."

Over the following decade, Freud metamorphosed from complacent researcher, pursuing his intuitions in a hospital laboratory in Vienna, to insecure clinician, struggling to establish a private medical practice so he could support a family, to some uneasy and perhaps unwieldy hybrid of the two, splitting his time between the laboratory and the clinic—between the theory and the practice of medicine. Yet even in private practice Freud continued to monitor neuroanatomical research as it raced toward its seemingly inevitable conclusion. If each central nerve cell in the brain exists in isolation from every other central nerve cell, as researchers had determined within Freud's lifetime, it must still establish a connection with other central nerve cells. So: Where was it? Find that specific point of connection, as the neuroanatomists of Freud's generation understood, and you'll have found, at long last, the final piece in the puzzle of man.

As recently as the first quarter of the nineteenth century, that puzzle had seemed potentially insoluble, in large part because of the limitations inherent in the only instrument that might conceivably assist such an investigation. Back in the 1670s, when Antonius von Leeuwenhoek and other natural philosophers began reporting what they could see by examining terrestrial objects through a microscope, this new method of investigation might have seemed to offer limitless possibilities for anatomists. That promise, however, pretty much vanished once they got a good look at the infinitesimal scale of what they'd be studying. Leeuwenhoek himself reported that the finest pieces of matter he could see in animal tissue were simply "globules," and for well over a century anatomists were left to concoct hypotheses about what these globulous objects might be without being able to see them any better than Leeuwenhoek had. As late as 1821, the English surgeon Charles Bell—who himself had only just recently distinguished between those nerves that carry the sensory impulses, or sensations, *to* the brain and those that carry the motor impulses, or instructions on how the body should respond, *from* the brain—gave up on the brain itself: "The endless confusion of the subject induces the physician, instead of taking the nervous system as the secure ground of his practice, to dismiss it from his course of study, as a subject presenting too great irregularity for legitimate investigation or reliance."

Only five years later, however, British physicist Joseph Jackson Lister effectively revolutionized the microscope through improvements to the objective lens—the one nearer the specimen—that mostly eliminated distortions and color aberrations. Before this advance, neuroanatomy hadn't been much more than speculation supported by inadequate, incomplete or imprecise observation—supported by further speculation even. Now a new breed of anatomist could explore tissue at a level of detail that was literally microscopic—the technique that would become known as histology.

In 1827, only a year after inventing his achromatic microscope, Lister himself (along with the physician Thomas Hodgkin) decisively refuted all the variations on Leeuwenhoek's "globular" hypothesis that had arisen over the preceding century and a half. The globules, it turned out, were illusions, mere tricks of the light that Lister's new lens system could now correct. By utilizing the achromatic microscope to examine brain tissue, Lister reported, "one sees instead of globules a multitude of very small particles, which are most irregular in shape and size." To these particles anatomists applied a term that the first microscopists back in the seventeenth century had used to describe some of the smallest features they could see, though now it came to refer only to the basic units of organisms, both plant and animals: cells. These cells, moreover, seemed always to be accompanied by long strands, or fibers. That a relationship between these cells and fibers existed in the central nervous system—the brain and spinal cord—formed the basis of Theodor Schwann's cell theory of 1839, a tremendous advance in knowledge from even fifteen years earlier. "However," as one researcher reported in 1842, "the arrangement of those parts in relation to each other is still completely unknown."

Some researchers argued that the fibers merely encircled the cells without having an anatomical connection to them. Some argued that the fibers actually emanated from the cells. Such was the density of the mass of material in the central nervous system, however, that even the achromatic microscope couldn't penetrate it sensibly. Despite the promise of clarity that Lister's improvements at first had seemed to offer, by the 1850s anatomists were beginning to resign themselves to the realization that the precise nature of the relationship between cells and fibers would have to remain a mystery, unless another technological advance came along.

It did, in 1858, two years after Freud was born, when the German histologist Joseph von Gerlach invented microscopic staining—a way to dye a sample so that the object under observation would bloom into a rich color against the background. The object under observation in this case was a central nerve cell, complete with an attendant network of fibers. Now neuroanatomists could see for themselves that the fibers and the cells were indeed anatomically connected. They could also see that the central nerve cells exist only in the gray matter of the brain and spinal cord, never the white. And they could see that even where a central nerve cell is part of a dense concentration within the gray matter, it exists apart from any other central nerve cell, in seeming isolation.

But if the cells themselves don't come into contact with other cells, then how do they communicate with one another, as they clearly must? How does one cell "know" what the others are doing and therefore act in concert and thereby register a sensation or commit the nervous system to a response, an action, a thought? If the answer wasn't in the cells, those solitary hubs, then it had to be elsewhere.

And the only elsewhere there was, was the fibers. Thanks to Gerlach's staining method, neuroanatomists now found that the fibers extend from the central nerve cells only into the white matter of the brain and spinal cord, never the gray. But there the trail went cold. The meshwork was still too intricate for anyone to trace the paths of all the fibers from a single cell to the point where the fibers terminate. And so yet another technical innovation needed to be invented, and so yet another one was. In 1873—the same year that Freud entered the University of Vienna as a medical student—Camillo Golgi, an Italian physician, developed a superior staining method that in effect isolated the fibers in the same way that Gerlach's had isolated the cells. Even so, researchers still couldn't find one point of connection between the fibers of two different cells. Golgi himself thought he found one in the 1880s, but his sample was inconclusive. Still, in order to do what central nerve cells do, which is pass along impulses to one another, both prevailing theory and common sense dictated that the fibers of neighboring cells *must* connect, somewhere.

Not until 1889 did the Spanish histologist Santiago Ramón y Cajal discover the truth: They don't, anywhere. *This,* then, was the basic unit of the brain, what the German anatomist Wilhelm Waldeyer two years later would name the neuron: each central nerve cell and its own fibers, existing apart from—that is, *not* connecting to—any other central nerve cells and their own fibers. But if even the least wisp of a fiber—a fibril—doesn't connect, what does it do? It *contacts,* Ramón y Cajal explained. It reaches out, under the excitation of an impulse, to touch the neighboring fibril or cell, and then, when the excitation has relaxed, it retracts to its previous state of isolation. "A connection with a fiber network," Waldeyer wrote, "or an origin from such a network, does not take place."

Although at first the "neuron doctrine," as Waldeyer christened it, might have seemed to contradict common sense, upon reflection the idea that communication between individual neurons was not continuous but intermittent actually went a long way toward a possible explanation of several otherwise inexplicable mental phenomena, such as the isolation of ideas, the creation of new associations, the temporary inability to remember a familiar fact, the confusion of memories. These were the phenomena, anyway, that Freud had been confronting in his private practice, where he found himself listening at length to hysterics, and wondering how to represent their worries and cures within the webwork of cells and fibers he remembered from his years of neuroanatomy.

"I am so deep in the 'Psychology for Neurologists' that it quite consumes me, till I have to break off overworked," Freud wrote to a friend in April 1895. "I have never been so intensely preoccupied by anything." By now Freud had embarked on a second career. From 1873 to 1885, first as a medical student and then as a medical researcher, he'd devoted himself to an examination of the nervous system—to research in neuroanatomy. In 1886, on the eve of his thirtieth birthday, he'd opened a private practice devoted to nervous disorders and, for the first time in his life, begun seeing patients, though he continued to conduct research on the side. After the development of the neuron theory, Freud would have had every reason to believe that if anyone were in a position to unite the psychical with the physical, it was he. He'd seen both sides. He'd studied both sides, immersing himself in the peculiar logic of each for long periods of time. He'd even written some tentative outlines to this effect over the past few years, in letters to his closest friend and constant correspondent, Wilhelm Fliess, a Berlin ear, nose, and throat doctor. Not until Freud could meet with Fliess personally late in the summer of 1895, however, and the two men could convene one of their days-long "congresses," as Freud liked to call these occasional periods of intense and inspirational professional discussion, did he see the project whole.

Freud began composing the manuscript on the train ride from Berlin back home to Vienna that September. "I am writing so little to you only because I am writing so much for you," Freud informed Fliess by letter on September 23. Barely two weeks later, on October 8, Freud mailed a draft to Fliess—a hundred or so handwritten pages in which he attempted to explain definitively the processes of the mind by describing exhaustively the mechanism of the brain that encases it.

And a mechanism it was. "The project," Freud wrote, in the second sentence of the manuscript, "involves two principal ideas": in essence, and in accord with Descartes's philosophy and Newton's physics, motion and matter. Freud's principal idea number 1 was straightforward enough: that the workings of the brain are "subject to the general laws of motion"—that matter moves immediately adjacent matter with comprehensive cause-and-effect predictability. What contributed to Freud's sense of urgency in composing this draft, however, was that he now knew what the "matter" was: "2. That it is to be assumed that the material particles in question are the neurons." He based this assumption, as he made explicit several pages later, on "the knowledge of neurons which has been arrived at by modern histology."

Yet even as he was passing the manuscript along to Fliess, Freud was starting to have his doubts. "I have been alternately proud and overjoyed and ashamed and miserable—until now, after an excess of mental torment, I apathetically tell myself: it does not yet, perhaps never will, hang together," he wrote in the accompanying letter. "I am not succeeding with the mechanical elucidation; rather, I am inclined to listen to the quiet voice which tells me that my explanations are not yet adequate."

In the weeks to come that inner voice softened briefly, then hardened again. "During one industrious night last week," Freud wrote to Fliess on October 20, twelve days after posting the manuscript, "the barriers suddenly lifted, the veils dropped, and everything became transparent—from the details of the neuroses to the determinants of consciousness. Everything seemed to fall into place, the cogs meshed, I had the impression that the thing now really was a machine that shortly would function on its own." On November 8, however, he reported that after other professional commitments had forced him to put the manuscript aside, he found he couldn't stop thinking about it—specifically, he noted with regret, that "it required a lot of revision. At that moment," he went on, "I rebelled against my tyrant. I felt overworked, irritated, confused, and incapable of mastering it all. So I flung it all aside. If you felt called on to form an opinion of those few sheets of paper that would justify my cry of joy at my victory, I am sorry, because you must have found it difficult." Freud added that in another two months, after he'd fulfilled his obligations, "I may be able to get the whole thing clearer." It was not to be. Only three weeks later he wrote to Fliess, "I no longer understand the state of mind in which I hatched the psychology; cannot conceive how I could have inflicted it on you. I believe you are still too polite; to me it appears to have been a kind of madness."

Maybe so. Whatever it was, it was over now, as if a fever had broken. The problem that Freud had found himself confronting was larger than pathways of nerves, larger than the neuron itself—or, maybe, smaller. Either way, it was the same problem that had been haunting physiology since the inception of the modern era more than two centuries earlier: brain. To be precise, it was brain in opposition to what the motions of matter within the human cranium represent: mind, maybe.

For much of human history, such a distinction would have been secondary, at best. The far more important distinction, instead, would have been the one between two types of matter: terrestrial and celestial. Down here, as Aristotle had said, were the four elements—earth, air, fire, and water, either alone or in any number of combinations. Up there was one element—quintessence, a single perfect substance that constituted the moon, sun, planets and stars, as well as the spheres that carried them on their heavenly journeys. An Earth that itself traveled through the heavens, however, not only erased the crucial distinction between what was terrestrial and what was celestial but—as Descartes appreciated when he was merely a budding philosopher—presented a strong argument that everything in heaven and everything on Earth might ultimately consist of the same stuff.

Descartes first heard about Galileo's discovery of four moons orbiting Jupiter in 1610, as a student at the Jesuit college at La Flèche. Although Descartes was only thirteen or fourteen when this astonishing news reached his outpost in the French countryside, he understood at once the profound effects such a discovery could have on philosophy and physics. The very scope of those effects, however, also reinforced for him two growing suspicions: that although philosophy "has been cultivated for many centuries by the most excellent minds," as he later wrote, "there is still no point in it which is not disputed and hence doubtful"; and that as "for the other sciences, in so far as they borrow their principles from philosophy I decided that nothing solid could have been built upon such shaky foundations." The only rational approach to this appalling and ongoing state of ignorance, he concluded, was to begin again, from the beginning—"to demolish everything completely and start right again from the foundations," and to do so by seeking "no knowledge other than that which could be found in myself or else in the great book of the world."

The World, in fact, was what he called his first attempt to explain all of physics. As was often the case with Descartes when he produced a work of physics, he simultaneously produced a companion work on how a reconception of physics would necessitate a new interpretation of man's role in it—a new physiology. This work he called *Treatise on Man.* He completed both in 1633, a year after Galileo released his own attempt at a new physics, *Dialogue Concerning the Two Chief World Systems.* But before his two volumes could reach publication, Descartes heard that the Roman Catholic Church had condemned Galileo because the *Dialogue* posited a sun-centered universe. Since his own two essays did the same and since he feared that if he altered them in any way they would be "mangled," Descartes suppressed both.

But he never stopped working on physics and physiology. In particular, over the next few years, he wondered if the new-found conceptual unity between the heavens and the Earth would allow him to achieve a parallel mathematical unity. In other words, could he do to the terrestrial realm what astronomers had long done to the celestial realm: geometricize it? Geometry, after all, had originally been an attempt to render the terrestrial world in mathematical terms. Now, after a lapse of a couple of millennia, it was again, and in his 1637 *Geometry,* Descartes demonstrated how all matter, not only in heaven but on Earth, could be located according to three coordinates in space. In which case, as Descartes himself recognized and as succeeding generations came to appreciate, a crucial question presented itself: Could we approach the secrets of man's inner universe with the same heretofore unthinkable curiosity that Galileo and *his* successors had regarded the outer? Could we render the motions of matter within the brain as predictable as any planet's through the heavens? In short, could there be a Newton of neurology?

Even while Newton was alive and evidence had begun to accumulate that his laws extend to the outermost reaches of the universe, the question inevitably had arisen whether those same laws might extend to the innermost reaches as well. In 1725, Richard Mead, an English physician, had produced mathematical formulations of the effects of planetary gravity on the human body. Expanding on that idea later in the eighteenth century, the German physician Franz Anton Mesmer proposed a gravitational attraction between animals, or what he called animal magnetism, whose existence he then claimed to demonstrate through public displays of hypnotism. In the early nineteenth century, efforts at quantifying psychic phenomena found a champion in the German philosopher Johann Friedrich Herbart, who conceived of the workings of the mind as "forces" rather than ideas, who explicitly invoked Newton in advocating the use of mathematical formulas to describe the motions of these forces, and who once declared, "Regular order in the human mind is wholly similar to that in the starry sky."

In retrospect, though, any such earlier efforts to reduce the workings of the inner universe to a series of cause-and-effect laws were doomed. These would-be Newtons couldn't have known it at the time, but they didn't yet have access to a Galilean equivalent of neuroanatomical data—the moons, planets, and stars of the inner universe—to provide their speculations with a solid empirical foundation.

Did Freud? It was tempting for him to think so. It would have been tempting for anyone in his position to think so—not only because it's always tempting for an ambitious intellect to think that the generation into which it's fortunate enough to be born is the one in possession of just enough information to settle a question that has thwarted the great thinkers since antiquity but because the state of neuroanatomical knowledge at the close of the nineteenth century *was* different from any other period in the history of science. In fact, in 1894—only five years after Ramón y Cajal's discovery that fibers from central nerve cells contact, not connect, and only three years after Waldeyer developed the neuron theory—one of Freud's former instructors and colleagues from his laboratory days, Sigmund Exner, published his own attempt at a comprehensive neuroanatomy, *Entwurf zu einer physiologischen Erklärung der psychischen Erscheinungen* (*Draft Toward a Physiological Explanation of the Psychological Features*).

Like most physiologists of his era, Freud knew firsthand what the achromatic microscope could accomplish. He'd used the still-new instrument extensively as a student in the 1870s, then proved his mastery of it the following decade as a reliable, respected diagnostician at the General Hospital of Vienna, where one of his examinations drew praise in a contemporary medical journal for its "very valuable contribution" to a field "heretofore lacking in detailed microscopic examination." And like many physiologists of his era, Freud knew firsthand what staining a microscopic sample could accomplish. He'd twice developed his own significant improvements on existing staining methods, first in 1877 "for the purpose of preparing in a guaranteed and easy way the central and peripheral nervous system of the higher vertebrate (mice, rabbits, cattle)," and again in 1883 "for the study of nerve tracts in the brain and spinal cord." And like a few physiologists of his era, Freud had even anticipated the neuron theory itself, during his lecture before the Vienna Psychiatric Society in the early 1880s, several years before Ramón y Cajal proved it. Unable to locate a fiber that he could trace from one central nerve cell to another, he'd wondered if cells might therefore *not* ultimately connect.

In the wake of his failure with the "Psychology for Neurologists," however, Freud began to consider another way to frame the problem: not as mind in opposition to brain—or at least not *only* mind in opposition to brain. Instead he began to think in terms of mind in opposition to *itself.*

"The starting-point for this investigation," Freud later wrote, outlining his reasoning at this juncture, "is provided by a fact without parallel, which defies all explanation or description—the fact of consciousness." On the most basic level, the workings of the mind remained a mystery. Even a thought, the fundamental unit of mind, doesn't remain in consciousness for any length of time. "A conception—or any other psychical element—which is now *present* to my consciousness may become *absent* the next moment, and may become *present again,* after an interval, unchanged." Forget for the moment the gap within the brain—between one neuron and the next, that space across which some "quantity" of "energy" must pass, as he'd tried to express the transaction in his "Psychology." And forget, too, the gap between brain and mind—between the physical communication among neurons and the resulting psychical impressions. With this description of one of the most mundane of human occurrences—something out of "our most daily personal experience"—Freud had identified a gap within the mind itself: "In the interval the idea was—we do not know what."

"Unconscious," he called it, adopting the common adjective of the time. In a sense, all he'd done was work his way back to the assumptions that he and his contemporaries had inherited. Mind was mind, brain was brain, and one day, maybe, the two would meet. *Brain* anyone with the proper training and equipment could tease the secrets out of, slicing tissue, staining samples, subjecting fibrils to microscopic scrutiny at recently unthinkable powers of magnification and degrees of resolution. *Mind,* however, nobody could fully capture using a mechanical model of the brain—not yet, anyway. Mind, as Freud could observe for himself on an almost daily basis in his private medical practice, was simply full of too many subtleties whose precise nature continued to doom any attempt to do for the physical workings of the inner universe what Newton had done for the outer.

But, in another sense, what Freud had learned through the experience of writing that manuscript was just how subtle the subtleties of mind were. Nothing in his neurological training had prepared him for—or, as he had now learned the hard way, could account for—that. Under intensive scrutiny, the mind had turned out to be even more complicated—far more circuitous, far more contradictory, and, finally, far more elusive—than he or, as far as he knew, anyone else had begun to imagine. Brain might be simply brain, but mind *wasn't* just mind.

As he reclined in a chair in his modest study in Vienna, listening to the complaints of patients week after week, year after year, Freud had learned to encourage them to try to see whether they could remember the trauma that had caused their hysterical symptoms. If they did so, as he tried to reassure them, their symptoms would disappear. Freud had first heard about this method many years earlier, back in 1882, from a friend and colleague in Vienna, the eminent physician and medical researcher Josef Breuer. At that time, Breuer had told Freud about how he'd treated a young woman's hysteria through hypnosis. Freud, in fact, had seen a demonstration of hypnosis once. It was, he thought, impressive, especially for a student with a physiological turn of mind. But instructive? *Curative?*

It might be so, said Breuer. Rather than simply issuing a

command or a prohibition while she was under hypnosis, he said, he had asked this patient—Anna O., Freud named her later, when recalling this period in his professional development—what the source of the trauma was. In her waking state she "could describe only very imperfectly or not at all" the memories relating to her trauma, as Freud later wrote; in a hypnotic state, however, she seemed oddly able to remember everything. Even more improbably, by recalling the source of the trauma, as well as by experiencing the emotional outpouring that invariably accompanied this memory, she seemed somehow to slip free of the grip of the memory—seemed to achieve, in Breuer's term, a "catharsis."

In order for this construct of a cure to hold, it might seem, the mind must work like a Newtonian machine: an initial cause leading to an effect, which in turn becomes a cause for another effect, which in turn becomes another cause for a further effect, and so on, insinuating itself throughout the subject's life until one day, in an unrecognizable guise, it surfaces as hysteric behavior, those worrisome symptoms that prompt the victim to seek medical attention. But if this description of the process were true, then the removal of any link along the way would be sufficient to interrupt the chain of causality and lead to the removal of the ultimate effect—the hysterical symptom. In that case, using hypnosis in a purely suggestive way, by simply commanding the symptom to disappear, would be sufficient in effecting a cure.

Freud, however, believed that he'd seen otherwise. When he attempted to apply this kind of therapy in his own medical practice, he found that leading the patient back to some step between the current state of hysteria and the original inciting incident didn't have a cathartic result. Only by revisiting the scene of the crime, so to speak, could a victim permanently break free of its memory, its insidious influence. Only by tracing it to its source would doctor and patient see the hysterical symptom disappear—but only by tracing it *all the way* to its source. As Freud told a meeting of the Vienna Medical Club in January 1893, "The moment at which the physician finds out the occasion when the symptom first appeared and the reason for its appearance is also the moment at which the symptom vanishes."

Anna O., for instance, complained of a paralysis on her right side, persistent hallucinations of snakes in her hair, and a sudden inability to speak her native German. These symptoms Breuer eventually traced back to an evening when she was nursing her sick father and imagined a snake approaching his sleeping figure. She tried to move to save him, but her right arm had gone to sleep over the back of a chair; and so she resorted to prayer, but in her fear all she could recall were some children's verses in English. Or, from Freud's own case files, Frau Cäcilie M., who suffered from a pain between her eyes until she remembered the time her grandmother had fixed her with a "piercing" look. "*Cessante causa cessat effectus,*" as Freud said in that same lecture before the Vienna Medical Club: "When the cause ceases, the effect ceases."

Freud, however, wasn't content with a vision of the mind that began with a cause and then, no matter what, ended with a certain effect. How to account for the inability of a process so powerful—so *active,* after all—to reveal itself?

With his background in a physiology that was ultimately nothing more than matter and motion, Freud knew exactly how to account for it: by postulating the existence within the unconscious of an opposing force at least equally powerful—a "defense" or, as Freud soon came to call it, a "repression," a change in terminology that itself reflected a change in Freud's thinking. This opposing force wasn't merely defending the mind against itself; it was repressing the unpleasant memory or association. It wasn't *re*active. It, too, was *active,* even while seemingly absent.

On October 26, 1896, Freud's father died. The heroic figure of Sigmund's childhood imagination may have disappeared forever during that long-ago walk when the father confided in the son how he'd submitted to the indignities of an anti-Semite, and now the corporeal figure was gone, too. Yet they lingered—both the heroic figure and the tragic shade. Like a traumatic event that remains present in the symptoms of a hysterical patient, the older man remained alive in the grown son. That night, in fact, Freud had a dream about him. On his way to the funeral, Freud stops at a barbershop. There he sees a sign: "You are requested to close the eyes." Whose eyes? he had to wonder. The dead father's? The son's? And "close" them as in lay to rest? Or "close" them as in "wink at" or "overlook"?

The dead-but-not-gone father wasn't the only thing that lingered. The dream did, too, taunting Freud with its myriad possible interpretations, haunting him like the earlier memory of the about-to-be-unheroic man on the street, inhabiting him, continuing to exert its influence over him, as an adult, decades after the event. In years to come, Freud more formally commemorated his father's death as "the most important event, the most poignant loss, of a man's life." But now, when his impressions were raw, he confided in a letter to his friend Fliess, "By one of those dark pathways behind the official consciousness the old man's death has affected me deeply."

Could Freud navigate those dark pathways? When he tried

to map the pathways of nerves within the brain, he had failed—and now he suspected it was because he'd set himself the wrong challenge. Now a new and radically different challenge presented itself to him: How to map the pathways of *the mind alone?* Even if he could, would anyone believe that such a description bears any resemblance to reality? *He* would, of course—but then, sitting in his office, listening to his patients, Sigmund Freud had heard the evidence for himself, if only in his mind's ear.

THREE

GOING TO EXTREMES

Now here was something nobody had ever seen before. The photograph that began appearing on the front pages of newspapers around the world in the first weeks of 1896 showed an image of a hand, more or less. Less, because this hand seemed to lack skin—or at least its outer layers of flesh and blood and tendons had been reduced to a presence sufficiently shadowy so as to allow a look beneath them. And more, because of what that look within revealed: the intricate webwork of bones that previously had been solely the province of the anatomist.

The hand belonged to the wife of Wilhelm Conrad Röntgen, a professor at the University of Würzburg in Germany. On November 8, 1895, while working alone in his darkened laboratory, Professor Röntgen had noticed a seemingly inexplicable glow. On closer inspection, this glow revealed mysterious properties. For the next several weeks Röntgen worked in secrecy, strictly adhering to the method that had initiated the Scientific Revolution more than two centuries earlier and had sustained it ever since: He made sure that anyone else could use a Hittorf-Crookes tube and a Ruhmkorff coil to produce and reproduce the effect he'd detected. Sometimes his wife, Bertha, would ask why he was spending so much time in his laboratory, and he would answer that he was working on something that, if word got out, would have people saying, *"Der Röntgen ist wohl verruckt geworden"* ("Röntgen has probably gone crazy").

At last Röntgen satisfied himself that his discovery was legitimate—that he hadn't somehow misinterpreted the data. On December 22, he invited Bertha to join him in the laboratory, where he asked her to insert her hand, for fifteen minutes, between the tube and a photographic plate. As she did so, something happened—something he'd witnessed for himself numerous times now, something he still couldn't explain. A substance passed between the tube and the plate—*must* have passed, because even though the substance itself was invisible, the effect was undeniable. An image of his wife's hand, to her horror, was slowly burning itself into existence. On New Year's Day, Röntgen went for a walk with Bertha, during which he mailed to colleagues copies of this photograph as well as his preliminary report, *"Eine neue Art von Strahlen"* ("A New Kind of Ray")—what Röntgen, in a footnote, christened "X rays," because of their mysterious nature. On the way to the mailbox, Röntgen turned to Bertha, whose hand—more or less, and complete (so to speak) with wedding ring—would soon be immortalized, and said, "Now the devil will have to be paid."

The first public news account appeared on January 5, after a professor of physics at the University of Vienna received a copy of the paper and photograph and passed them along to colleagues, who in turn contacted the editor of the *Wiener*

Presse. From there the news spread rapidly: the *London Daily Chronicle* on January 6, the *Frankfurter Zeitung* on January 7. "Men of science in this city," began an article in *The New York Times* a few days later, "are awaiting with the utmost impatience the arrival of European technical journals"; when an English translation of Röntgen's report arrived, the *Times* printed it on the front page almost in its entirety. By January 13 Röntgen had been summoned to Berlin, where he gave a demonstration of X-rays before Kaiser Wilhelm II. Röntgen granted only one substantive interview, to a particularly enterprising journalist from *McClure's Magazine,* then withdrew from public scrutiny. "In a few days I was disgusted with the whole thing," he later recalled. "I could not recognize my own work in the reports anymore."

It was entirely coincidence that Röntgen had conducted what proved to be his decisive research at the same historical moment, and possibly even the same literal moment, that the sixteen-year-old Albert Einstein was pacing the grounds of a school in Aarau, trying to reconcile the properties of a beam of light with the concept of absolute space. It was similarly a coincidence that Röntgen made his discovery on the very day that the thirty-nine-year-old Sigmund Freud was writing to his friend Wilhelm Fliess that he was thinking of abandoning his attempt to render human thoughts strictly in terms of the motions of neuroanatomic matter. But it was no coincidence that speculation about the implications of Röntgen's discovery led in the same direction that Einstein and Freud were already heading.

If X-rays were a "longitudinal vibration in the ether," Thomas Edison pointed out to a journalist visiting his New Jersey laboratory, "then Professor Röntgen has found out at least one method of investigating [the ether's] properties, and the gain to men of science in estimating the behavior of light and electricity through the medium of ether will probably be immense, causing many changes in our present theories." Meanwhile, *Science* magazine reported that the College of Physicians and Surgeons in New York City had used the rays "to reflect anatomic diagrams directly into the brains of advanced medical students, making a much more enduring impression than ordinary teaching methods of learning anatomic details." And whatever went into a skull presumably could also come out: A Mr. Ingles Rogers informed a San Francisco newspaper that he had "produced an impression on the photographic plate by simply gazing at it in the dark," while a Dr. Baraduc attracted worldwide attention by sending an official communication to the Paris Académie de Médecin that he had succeeded in photographing thoughts, and he mounted an exhibition in Munich as proof. Or, as one early account regarding X-rays cautioned readers who would fool themselves into thinking that if they "go inside the house and pull down the blinds and wait till it is dark," they might "feel quite safe in sinning": "There are the x-rays, you know—and nobody knows what other invisible pencils may be registering all our actions or even thoughts—or what's worse, the desires that we don't dare think. They, too, must leave their mark somewhere."

Nobody knew quite what X-rays were—not even Röntgen, though he did tentatively guess that they might indeed be longitudinal vibrations in the ether. And nobody knew quite what X-rays did—though the medical applications were apparent enough that by the following autumn *The New York Times* reported "no hospital in the land can do justice to its patients if it does not possess a complete X-ray outfit." And nobody could have known what impact X-rays eventually would have on the history of science—revolutionary, but that's getting ahead of the story. What anybody who cared about science did know by now, though, following more than two centuries of investigation, was that in the properties of the ether or the pathways of the brain the scientific method had reached the frontiers of the outer and inner universes.

In which case: Was the store of knowledge finite? And if so, was the history of science now nearing its end?

The idea of science's even having a history was a novel concept. *History,* in its modern interpretation, was still a fairly recent addition to Western thought. Before the dawn of the modern era of science, time had seemed—to the extent it had seemed anything—unchanging or, at most, cyclical. People lived and died; civilizations rose and fell. The particular circumstances of such a world might change, but not its essence.

Not so the essence of a world that placed a premium on the acquisition of new knowledge—on change itself. When Western scholars of the fifteenth century began resurrecting texts from antiquity and rediscovering ancient knowledge, they would have had no reason to believe that these sources were inaccurate or in any way incomplete. On the contrary: They had every reason to assume the texts they were inheriting were correct and comprehensive. Didn't these writings, whether in the original Greek or in Arabic translation, represent the cumulative knowledge of a civilization at the zenith of its intellectual powers? True, that particular civilization had risen, and it had fallen. And true, civilization itself had lain dormant for the following thousand years. But the essence of civilization itself didn't change: the understanding of the way the world works.

Yet that's what the understanding of the way the world

works was doing now: changing. Ptolemy of Alexandria, writing in the second century A.D., had compiled such a seemingly thorough catalogue of the workings of the heavens that some seven centuries later his Arabic translators honored his work with the title of "The Greatest Composition," or *Al-mageste* (later shortened to *Almagest*). Galen of Pergamum, also writing in the second century A.D., had described the human anatomy in more than a hundred texts that survived the ages, a volume of output that alone must have intimidated succeeding generations into accepting his word as gospel on all matters medical. Yet in 1610, Galileo looked through a telescope and determined to his own satisfaction that much of what Ptolemy had written was inaccurate, while in the 1630s Descartes surveyed the current state of medical knowledge and concluded, "I am confident that there is no one, even among those whose profession it is, who does not admit that all at present known in it is almost nothing in comparison of what remains to be discovered."

So the acquisition of knowledge was only now just beginning, right? Wrong. Just ask Galileo. "It was granted to me alone to discover all the new phenomena in the sky and nothing to nobody else," he wrote of what he'd observed through the telescope. Such was Galileo's authority that his assessment was echoed, a generation after his death, by no less an eminence than Christopher Wren: "All celestial mysteries were at once disclosed to him. His successors are envious because they believe that there can scarcely be any new worlds left."

Galileo, for once, had missed the point. No doubt his legendary arrogance contributed to his high opinion of his own accomplishments. So did the indisputably monumental and lasting nature of his additions to the astronomical and intellectual landscape. In the absence of a precedent for a single life's work such as his, it might very well have seemed what Galileo assumed it to be: a self-standing anomaly, a onetime correction to the core of human knowledge—even, perhaps, its completion.

Put enough anomalies together, however, and that interpretation of a life's work begins to shade in the direction that Descartes indicated. By the time Newton wrote in a letter to a friend in the 1670s that he had seen farther than his predecessors only because he was standing on the shoulders of giants, so many seemingly anomalous individuals had appeared in physics and astronomy that they were clearly no longer the exceptions but the rule. Through his choice of reference, Newton was indicating as much. He was paraphrasing the philosopher Bernard of Chartres, who in 1115 had written, "We are like dwarfs sitting on the shoulders of giants; hence we can see more and further than they, yet not by reason of the keenness of our vision, but because we have been raised aloft and are being carried by men of huge stature." The cathedral then just beginning to rise in Bernard's own village eventually commemorated this sentiment with lancet windows depicting New Testament preachers perching on the shoulders of Old Testament prophets, and it was this new interpretation of time that Newton, half a millennium later, was pointedly evoking: history as a cumulative enterprise, science as a cathedral of knowledge.

Even this shift in attitude regarding individual contributions, however, didn't fully capture why Galileo was wrong. In his estimation of the discoveries he'd made by following the modern method of interrogating nature—looking for himself—he had failed to appreciate the value of the further innovation that he himself had introduced to the scientific process: how he'd made the discoveries. He'd used and refined a new instrument, the *perspicillum,* or perspective tube, and as that original name for what was soon called the telescope might suggest, Galileo was no longer simply seeing what was out there, but seeing *more.*

His error was understandable. In the centuries leading up to the invention of the telescope, knowledge of optics had encouraged some fabulists to fantasize about an instrument that would make distant objects appear near. Some day they would see coins in distant meadows. They would see steeples in distant towns. They would see ships on distant horizons. And now, that day had arrived. The invention of the telescope—nothing more than a couple of discs of glass in a tube, really—allowed them to see everything they had imagined. They saw distant coins, steeples, ships, all right.

But who could have imagined all that Galileo discovered through the telescope: not currency, Christianity, or commerce, not more evidence of a previously, if imperfectly, explored world, but evidence of *more worlds*? Of moons around Jupiter, multitudes of stars, and imperfections on the surface of the sun and moon—the sun and moon! those two daily companions, presumably as familiar as the proverbial back of one's hand—that rendered them as good as new.

This pattern of anticipation and surprise repeated itself in the decades that followed Galileo's revolutionary investigations, as the success of the telescope as well as improvements to optics encouraged some natural philosophers to fantasize about an instrument that would make small objects appear large. And, indeed, nobody could have imagined the millions of "animalcules" that Leeuwenhoek, starting in 1674 and continuing for years to come, sighted in everyday objects such as, yes, the skin on the back of one's hand.

But already the larger meaning behind such discoveries had

changed. Leeuwenhoek's new universe was audacious, epoch-making even, but that wasn't the point—or at least not the only point that natural philosophers would have taken from this finding. This universe was impossible to see without a microscope, but that wasn't the point, either. In 1655, the Dutch astronomer Christian Huygens had used a telescope to find a world that Galileo hadn't—a moon of Saturn—and then, in 1659, he'd resolved a visual mystery that had stumped Galileo—that the odd and changing shape of the planet's appearance was due to, of all things, a ring. Discoveries of more moons followed, in 1671 and 1672; then, in 1675, a division in the ring itself. By the time the microscope came into its own as a means of investigating nature's secrets, discoveries such as Leeuwenhoek's, however impossible to anticipate, would have represented not the end but the beginning of a great intellectual adventure—and *that* was the point.

Galileo, Leeuwenhoek, and their intellectual descendants could see "further," as Bernard of Chartres had said and Newton echoed, not only because they were sitting on the shoulders of giants and not only because they were sitting on the shoulders of giants and looking for themselves but because they were sitting on the shoulders of giants and looking for themselves through instruments that allowed them to see *more* than meets "the naked eye"—a term that natural philosophers began to invoke only now, in the last half of the seventeenth century, to distinguish present discoveries from all preceding ones. And how much more could they see? This was the question that revealed the limits of Galileo's otherwise seemingly limitless imagination, and this was the question that had driven Descartes to reimagine the pursuit of science. Having gotten your first glimpse of the universe beyond what your senses alone can detect, you could assume, like Galileo, that that's all there was and then stop looking. Or you could wonder, like Descartes, what else might be out there and trust that something was.

Despite the absence of a precedent for the Scientific Revolution, there was at least a possible analogy—an almost exact historical parallel, in fact—the Age of Discovery. Like the ocean voyages that, starting in the early fifteenth century, had taken European explorers to Africa, Asia, and the Americas, the Scientific Revolution that began in the sixteenth century had opened the borders of similarly unexamined and unimagined lands. The parallel was hardly a coincidence; each had arisen out of the same impulse, a curiosity-unto-conquest. By the late nineteenth century, however, the geographical explorations had exhausted all but the most inaccessible interiors (of Africa, for instance) or the most remote outposts (the polar regions). "No one would claim for a moment that during the next five hundred years the accumulated stock of knowledge of geography will increase as it has during the last five hundred," a future president of the American Association for the Advancement of Science, T. C. Mendenhall, wrote in 1887, explicitly invoking this metaphor. "More than ever before in the history of science and invention, it is safe to say what is possible and what is impossible."

Were there still more things in heaven and earth? Increasingly, the answer was: Maybe not. As the British physicist William Dampier later recalled of his apprenticeship at Cambridge during this era, "It seemed as though the main framework had been put together once for all, and that little remained to be done but to measure physical constants to the increased accuracy represented by another decimal place." In fact, the "sixth place of decimals," in the words of the American physicist (and would-be ether-drift detector) Albert A. Michelson, became a common shorthand for the arena in which future investigators of nature would have to operate.

How one regarded this redefinition of the scientific enterprise depended on what one believed this narrower approach could accomplish. For the British physicist Joseph John Thomson, a belief that "all that was left was to alter a decimal or two in some physical constant" evoked a "pessimistic feeling, not uncommon at that time, that all the interesting things had been discovered." As the premier American astronomer of the era, Simon Newcomb, wrote in 1888, "If the public are disappointed at not seeing brilliant discoveries coming from all the observatories, we must remember that what are called discoveries are not the main work of the scientific man of today." Instead, the main work was filling in the details of what had already been discovered, and the reason, Newcomb said, was simple, at least in astronomy: "it must be confessed that we do appear to be fast approaching the limits of our knowledge."

For some scientists, however, a narrower focus was actually a cause for optimism. As early as 1871, the physicist James Clerk Maxwell, in his inaugural address as Cambridge University's first Cavendish Professor, expressed his dismay "that the opinion seems to have got abroad, that in a few years all the great physical constants will have been approximately estimated, and that the only occupation which will then be left to men of science will be to carry on these measurements to another place of decimals." Such pessimism, he thought, got the scientific method precisely backward. Science cannot advance, he emphasized, without "improving the accuracy of the numerical measurement of quantities with which she has long been familiar." As William Thomson, the future Lord Kelvin, wrote in *Nature* during this same period, "Accurate and minute measurement seems to the non-scientific imagination a less lofty and dignified work than looking for something new. But nearly all the grandest discoveries of science have been but the rewards of accurate measurement and patient long-continued labour in the minute sifting of numerical results." Michelson himself, regretting the negative interpretation that had attached to his earlier invocation of the sixth place of decimals while doubtlessly also drawing on his experience with the seemingly not-quite-sensitive-enough ether-detection interferometers in Berlin, Potsdam, and Cleveland, wrote in 1899, "A hundred years ago a measurement made to within one thousandth of an inch was considered rather phenomenal. Now it is one of the modern requirements in the most accurate machine work. At present a few measurements are relied upon to within one ten-thousandth of an inch. There are cases in which an accuracy of one-millionth of an inch has to be attained and it is even possible to detect differences of one five-millionth of an inch. Past experience indicates that we are merely anticipating the requirements of the not-too-distant future in producing means for the determination of such small quantities."

At least part of the problem in trying to address the question of completeness at the turn of the twentieth century was virtually the same as the one at the turn of the seventeenth century. Back then, it was that nobody in history had ever lived through a period like the one that might or might not be beginning. Now it was that nobody in history had ever lived through a period like the one that might or might not be drawing to a close. Never before had the facts of nature, as well as whatever overarching truths they might suggest, been pursued so avidly by so many for so long, and as a result there was no way of knowing whether any new understanding of the universe that happened to emerge would turn out to be the final one or whether future discoveries would offer further refinements of current knowledge. If natural philosophers in 1600 lacked precedents for individual discoveries of the kind that Galileo and Leeuwenhoek soon made, then natural philosophers in 1900 lacked precedents for the *collective* discoveries that physicists and physiologists had made over the preceding three centuries. Was the discovery of an apparent anomaly such as X-rays the key to completing one of the two most ambitious scientific programs in history? Or was it representative merely of the many ways that medical knowledge would continue to deepen? Either way, natural philosophers at the turn of the twentieth century could be sure of at least this much: By *looking,* and then looking some more, investigators of nature had narrowed their searches to previously, and recently, unfathomable levels of precision, and those new lands had turned out to be just as navigable—as conquerable—as the Americas.

Yet when Einstein and Freud each got his first glimpse of a new universe, it turned out to have nothing to do with finding a further place of decimals or exploring new numerical results. Another improvement in telescopes wasn't going to locate the ether. Another improvement in microscopes *had* located the neuron—and it wasn't enough. But each man, in his own way and dutifully following the example of all the giants on whose shoulders he sat, had looked, until he'd run out of evidence using the standard means of investigation. Then, because something still didn't make sense, he'd kept looking anyway, until he found it: not something farther, not something deeper—not something *more,* but something else—something *other.*

When Einstein began investigating the problem of the ether, he was content to follow the example of his contemporaries. First he tried to imagine instruments that would detect the ether wind. Then he concentrated his attentions on the question of absolute space. Even when he realized that the possible solution to the problem lay in a reconsideration not of absolute space but of absolute time, he was still covering the same material as his most radical peers.

Einstein always marveled that someone else didn't beat him to relativity—didn't reach the same insight into absolute time that he did on that night in Bern in May 1905. "The step," Einstein liked to call it in years to come, as if it had been nearly inevitable, merely a matter of putting one foot in front of another. As Einstein later noted in an obituary of Ernst Mach, the leading German-language philosopher of science at the time and one of the two greatest influences on how Einstein thought about physics: "It is not improbable that Mach would have discovered the theory of relativity, if, at the time when his mind was still young and susceptible, the problem of the constancy of the speed of light had been discussed among physicists."

One person, in fact, did arrive at virtually the same mathematical formulation for relativity as Einstein, and even a year earlier: Hendrik Antoon Lorentz, the physicist who Henri Poincaré felt had fashioned one too many hypotheses in attempting to explain the null result of Michelson's various ether-detection experiments. The fact that Einstein and Lorentz independently reached two versions of the same equations wasn't quite a coincidence; each set of figures arose out of

Maxwell's equations on electromagnetism in the 1860s and the idea of, as Lorentz summarized it in the title of his paper, "Electromagnetic Phenomena in a System Moving with Any Velocity Smaller Than That of Light." Still, it was because of the similarity between the two versions that Einstein's paper, upon its publication in the *Annalen der Physik* on September 28, 1905, attracted less attention for the propositions in its first pages than for the mathematics in its final: Many readers simply interpreted Einstein's version as a slight improvement over Lorentz's, and for several years the formula was known as the Lorentz-Einstein theory.

Only gradually did the differences between the two versions become apparent—not so much how the formula itself worked, but the reasoning that led to it, and the conclusions that followed from it. Lorentz had set out to describe the contraction of electrons (in Michelson's interferometer, for instance) under pressure from the ether as the Earth barrels through it. But to Einstein, because of the two postulates he introduced in the second paragraph of his paper—that Galileo's principle of relativity should extend to electrodynamics and optics and that the speed of light in a vacuum is constant—the ether was "superfluous." Then, Lorentz's formula described physical changes that bodies approaching the speed of light would undergo. But Einstein's described *perceptual* changes that *observers* would experience. It described measurements contingent on the relative motion of the object under observation, due to the high, finite—and, most important, constant—speed of the electromagnetic waves conveying the information to an observer.

The peculiar property of the constancy of the speed of light was one that Poincaré—probably the most eminent mathematician of the day and, with Mach, the other great influence on Einstein's thinking—had struggled with. Unlike Mach, who was a generation older, or Lorentz, who belonged to his own generation, Poincaré's interest in the question wasn't only theoretical. It was practical. In 1893 Poincaré had become a member of the Bureau des Longitudes in Paris; in 1902, a professor at the École Professionelle Supérieure des Postes et Télégraphes. In both capacities, he'd needed to attend to one of the most pressing logistical matters of the day: the coordination of clocks—specifically, the coordination of *electrical* clocks.

Unlike mechanical clocks, electrical clocks allowed the transmission of information from town to town, city to city, capital to capital, even shore to ship once radio signals came into use, and all at the speed of light. No longer need each locality base its own time on that moment when the sun passed overhead at noon. Now, entire longitudinal swaths of countries or continents could keep to the same 24-hour day schedule—a considerable improvement in terms of train arrivals and departures, military movements, maritime logistics, and other exceedingly practical matters. In the years immediately preceding and following the turn of the twentieth century—and nowhere more than in the unofficial capital of Europe, Paris, where Poincaré presided over virtually all concerns chronometric—clocks fell into lockstep formation.

It hadn't been easy. In fact, Poincaré in his writings and lectures frequently chose as his topic the difficulty of trying to arrive at a definition of simultaneity. In a widely read and influential 1898 essay, *"La Mesure du temps"* ("The Measure of Time"), he wrote that a geographer or navigator wanting to know the time in Paris without being in Paris could rely on the transmission of a telegraphic signal. "It is clear first that the reception of the signal at Berlin, for instance, is after the sending of this same signal from Paris," he wrote. "But how much after? In general, the duration of the transmission is neglected and the two events are regarded as simultaneous."

Poincaré understood as well as anyone that the speed of light acts as a natural limit in the transmission and reception of information. Two years later, in a 1900 essay, and then four years after that, in an address before the International Congress of Arts and Sciences at the World's Fair in St. Louis, Poincaré invited his audience to join him in imagining two observers as they try to synchronize their clocks through the use of light signals. "The clocks synchronized in that matter will not mark the true time," he said, "but what one might call 'local time,' so that one of them is slow with regard to the other. This does not matter much," he added, almost as an afterthought yet fatally, "as we have no way of determining it." This final comment recalls what the seventeenth-century English astronomer Robert Hooke had written on hearing that Ole Rømer had fixed the velocity of light by studying the eclipses of the moons of Jupiter: "It is so exceeding swift . . . why it may not be as well instantaneous I know no reason."

Hooke knew no reason for a good reason: He hadn't yet absorbed the demands of a new, emerging conception of the universe, one where such fractional considerations would assume greater and greater importance. The same was now true of Poincaré. Advances in electromagnetic technology had made clock coordination possible, but advances in other kinds of technology had made it *essential.* A train not following the tightest of schedules might (and sometimes did) find itself barreling straight toward a train approaching on the same track from the opposite direction. For all his investment in the philosophical, physical, and practical considerations of the issue,

Poincaré failed to appreciate fully the need to become, as the director of one of the most prominent telegraph concerns in Switzerland in this period said, "master not only of the hour but also of the minute, the second, and even in special cases the tenth, the hundredth, the thousandth, the millionth of a second."

Had Einstein absorbed the demands of a new, emerging conception of the universe more thoroughly than Poincaré? Perhaps. Almost certainly his opportunities to do so had been at least equally as favorable as those available to Poincaré. As part of Einstein's work in the patent office, he would have routinely come across applications for components of clock-coordination systems. And to bring the question home, in a quite literal sense, Einstein moved his family in May 1905 from a central section of Bern, where the clocks in the towers were tethered electronically to a "master clock," to an outlying neighborhood from which he could see the tower in the suburb of Muri, where the clocks were not. As Einstein later reported about his decisive moment, "An analysis of the concept of time was my solution."

So why Einstein and not Poincaré? Poincaré himself, though writing in a different context, once hinted at a possible reason. Reflecting on the scientific process in general, Poincaré wrote, "We must, for example, use language, and our language is necessarily steeped in preconceived ideas. Only they are unconscious preconceived ideas, which are a thousand times the most dangerous of all." When he came to the topic of time, this proved to be the obstacle he couldn't overcome. As Einstein later wrote, the illusion of absolute time—of simultaneity—"unrecognizably was anchored in the unconscious."

Consciousness, if Freud were correct, was unrecognizably anchored in the unconscious, too. "The overwhelming majority of philosophers," Freud once wrote, "regard as mental only the phenomena of consciousness. For them the world of consciousness coincides with the sphere of what is mental." This might have been news to the majority of philosophers who had been arguing since the days of Descartes that the concept of consciousness should not and could not encompass everything psychical (and Freud himself admitted that he sometimes wished he were better read in philosophy). If nothing else, though, Freud's description of the assumption he thought he was overturning probably represents a fair summary of the assumption he himself had held—the assumption he'd unknowingly brought to his efforts to trace the fibers and fibrils of the brain or, later, to map onto the brain the functions of the mind. And even when he concluded that the mind itself contained conflicts and contradictions, he was identifying nothing new.

The possible existence of thoughts outside of conscious experience had been recognized for as long as man had been thinking about thinking. The Greek physician Galen had written about it. Two hundred years later, and twelve hundred years before Descartes, St. Augustine asked how a thought could be outside memory, momentarily seemingly irretrievable, yet at the same time within memory, ultimately retrievable—absent yet present. If anything, Descartes's rigid distinction between matter and mind—and especially its implicit suggestion that if mind is an external entity, it is a single, *indivisible* external entity—served only to highlight this semantic gap, which generations of physiologists and philosophers then sought to fill. Even before they had a word for it, they settled on a metaphor: "Our clear concepts," Gottfried Wilhelm Leibniz wrote in the early eighteenth century, "are like islands which arise above the ocean of obscure ones." Forms of the word "unconscious" began entering the English language in the mid-eighteenth century, German (*Unbewusstsein* and *bewusstlos*) in the late eighteenth century, and French (*inconscient*) in the mid-nineteenth century.

If, as Poincaré said, our language is in fact steeped in preconceived ideas, then maybe the addition of a new word to the language suggests the arrival of a *re*conceived idea—or at least the beginnings of one. In the German language, at any rate, this seemed to be so. "There is in intelligence," the physician and psychologist Wilhelm Griesinger wrote in 1845, "an actual, though to us an unconscious, life and movement; we recognize it however by its results, which often suddenly make their appearance from some unexpected source. A constant activity reigns over this almost, if not wholly, darkened sphere, which is much greater and more characteristic for the individuality than the relatively small number of impressions which pass into the state of consciousness." One year later, the physiologist and psychologist Karl Gustav Carus wrote in the opening sentence of his enormously influential *Psyche,* "The key to the understanding of the character of the conscious life lies in the region of the unconscious." By 1869, the young German philosopher Eduard von Hartmann could identify in his *Philosophie des Unbewussten* (*Philosophy of the Unconscious*) at least twenty-six aspects of unconscious mental activity dating back to Descartes. Maybe more than any other work of the era, this massive book popularized the idea of an unconscious; its ideas and the name of its author entered the German vernacular during Freud's teen years. French and English translations quickly followed, as did many other books on the unconscious in all three languages. By 1890, the American William James was comparing consciousness to a stream.

Like Freud, these philosophers and psychologists found that by postulating the existence of an unconscious, they could account for everyday effects—creativity, will, emotions—that otherwise might seem to lack causes. Unlike Freud, however, they had no reason to think they'd witnessed the unconscious in action. They had seen effects, and they had postulated causes, but they hadn't witnessed for themselves causes *leading* to effects. More accurately, they hadn't witnessed—as Freud thought he had, in conducting the cathartic therapy recommended by his colleague Breuer—the removal of causes leading to the *removal of effects.*

Even then, Freud wasn't alone in staking a claim for cathartic therapy or in sensing its implications for the conception of the unconscious. When the lecture in which Freud had introduced the formulation "When the cause ceases, the effect ceases" appeared in print in January 1893, it bore the heading "By Dr. Josef Breuer and Dr. Sigm. Freud of Vienna." Freud, in fact, had adapted the lecture from "On the Psychical Mechanism of Hysterical Phenomena (Preliminary Communication)," the summary of his and Breuer's work that they published first that same month in medical journals in Berlin and Vienna and then again in 1895 as the opening chapter of a collaborative volume of case histories, *Studies on Hysteria.* The impetus for that book was Freud's fascination with Breuer's cathartic cure of Anna O., and together Breuer and Freud had developed the cause-and-effect conclusion that they first presented (at Freud's behest) in the "Preliminary Communication": "the determining process continues to operate in some way or other for years—not indirectly, through a chain of intermediate causal links, but as a *directly* releasing cause." Theirs, then, was a new conception of the unconscious, or *"das Unbewusste,"* the term that Breuer and Freud first invoked as a noun, rather than as an adjective (as in "unconscious memory"), in the *Studies on Hysteria*—not absent yet present, but this: absent yet active.

Even as Breuer and Freud were publishing their "Preliminary Communication," however, the reasoning that had led them to this reconception of the unconscious was diverging, and so too the conclusions that followed from it. To Breuer, an idea that was somehow absent yet active suggested a sort of predisposition toward a certain behavior—a "hypnoid state." And Freud agreed, at first. Yet even before collaborating with Breuer on the "Preliminary Communication," Freud had begun entertaining an alternative interpretation, one that could coexist with hypnoid states. In 1894, he published a paper delineating that alternative: "I shall call this form '*defense* hysteria,' using the name to distinguish it from *hypnoid* hysteria." The year after that, in the *Studies on Hysteria,* his ambivalence was beginning to show: "I willingly adhere to this hypothesis of there being a hypnoid hysteria. Strangely enough," he continued, "I have never in my own experience met with a genuine hypnoid hysteria." And the year after that, in a paper he developed from a lecture on hysteria, Freud could barely contain his impatience with the concept: "I find, however, that there are often no grounds whatever for presupposing the presence of such hypnoid states."

From Freud's point of view, part of the problem with the hypothesis of a hypnoid state was that it introduced a new term—and a need for a new explanation—without really explaining anything itself. As Freud later wrote, "This opened the further question of the origin of these hypnoid states." Freud's own conception of a defense, or a repression, *did* offer an explanation. Not only does the original trauma remain active within the unconscious for days or decades, but so too does the repressive force that renders the ongoing effects of that initial cause seemingly absent.

Whether or not the store of knowledge was finite, whether or not science was nearing its end, Einstein and Freud each knew that his own investigations were now only beginning. The promise of the ether had turned out to be a false one. The ether didn't exist—didn't even need to exist, in order to account for what Einstein thought he'd found on his perambulations around Bern. The promise of the neuron turned out to be false, too. The neuron did exist—but still couldn't account for what Freud thought he'd found in his conversations with patients in Vienna.

Of course, just because one person thought he'd found something new didn't mean he actually had—especially if what he'd found was something that, once word got out, might have people saying *Der Einstein* or *Der Freud ist wohl verruckt geworden*: Einstein or Freud has probably gone crazy. History was full of visionaries who thought they'd discovered a promised land, only to have their descriptions of what it was like match the experience of nobody else—only to have their findings fail to withstand the scrutiny of the scientific method. If a particular frontier—a relativistic outer universe, an unconscious inner universe—was going to be found scientifically sound, it was going to have to be subject to independent verification. There was going to have to be evidence in its favor, and that evidence was going to have to match the evidence that anyone else could produce and reproduce. Until then, that frontier would have to reside only where an Einstein or a

Freud had encountered it—in that inscrutable, notoriously unreliable yet ultimately indispensable scientific instrument, one man's mind.

Even so—and even while beginning to subject their discoveries to the subtleties of the scientific method—Einstein and Freud found themselves considering a more immediate question, one that any natural philosopher confronting data that nobody had ever confronted before would have to ask: Is there . . . more?

II

MATTER OVER MIND

FOUR

A LEAP OF FAITH

In his mind, he was falling. Falling; and not falling. Behind him was a roof, on which he had been resting a moment earlier, and before him the ground, on which he would be resting a moment later, and in that sense he was falling. But in another sense, he was already at rest. Throughout the universe, every scrap of matter was exerting its exquisitely predictable influence on him—most of all, of course, the Earth, toward which he happened to be arcing at the moment; but also the roof from which his trajectory was taking him; and, too, the moon, as surely as it was tugging on the tides; and the sun, that centerpiece, an immensity orchestrating the orbits of all the bodies within its system; even the stars at their unimaginable distances, as well as whatever else might lie beyond them. Yet add everything up, and what he felt in that moment of free fall was precisely . . . nothing.

It was something, this nothing. It was an absence, to be sure: a lack of groundedness, a momentary suspension that also applied to anything else that happened to be falling along with him—a scale, for instance. A moment earlier, resting on the roof, and a moment later, resting on the Earth, that scale supporting that man would register an unsurprising quantity, his weight. But between those points of departure and arrival, here in midfall, scale and man would be traveling together at the same rate. If the man somehow remained standing on the scale during his descent, he would weigh, throughout his passage, zero.

And for that reason this nothing was, more than anything, a presence: a relationship previously hidden from our senses: the effect of acceleration and the effect of gravitation canceling each other out. And it was the recognition of this essential equivalence—this something and something resulting in nothing—that stunned Albert Einstein, stopped him cold, pinned him to his chair.

This was, he later recalled, "the most fortunate thought of my life"—one that potentially rivaled Newton's meditation on apples and gravity. Gravity, in fact, was what Einstein had been daydreaming about when this imaginative flight came to him, or he to it, in the autumn of 1907. He had been sitting in the patent office in Bern, Switzerland, where he was still working as a technical expert, and he had been thinking about a paper he'd been asked to write by the editor of the *Jahrbuch der Radioaktivität und Elektronik* (*Yearbook of Radioactivity and Electronics*). The purpose of this paper was to summarize and elaborate on "On the Electrodynamics of Moving Bodies," the *Annalen der Physik* paper on relativity that he'd written two years earlier. That earlier paper, along with several others he'd composed during the same stretch in 1905, had advanced Einstein's reputation among physicists, though not yet the circumstances of his life. "I must confess to you," a fellow physicist wrote to Einstein during this period, preparing to travel the two hundred miles from Würzburg to Bern for the purpose of collaborating with this theorist whose papers he admired and responding to a letter from Einstein saying where he might be found, "that I was amazed to read that you have to sit in an office for eight hours a day! But history is full of bad jokes."

Einstein himself wasn't complaining. Twenty-eight years old, the father of a three-year-old son, Einstein enjoyed numerous advantages through his post at the patent office: a steady (even "handsome") income; his close friendship with Besso, the co-worker also interested in physics; a dependable schedule that, even though it consumed forty-eight hours of the week, still left "eight hours of fun in the day, and then there is also Sunday." Not that the schedule was ideal. Already Einstein had asked the editor of the *Yearbook* to forward some background literature to him, "as the library is closed during my free time." Still, Einstein appreciated the patent work itself—"enormously varied," he once described it in a letter to a friend, "and it calls for much thought." "Many-sided" thought, he specified on another occasion; a technical expert such as Einstein had to evaluate an inventor's application for a patent on the basis of not only originality but practicality, purely by examining drawings and specifications. "When you pick up an application," the head of the patent office had instructed him, "think that anything the inventor says is wrong." Otherwise, his boss went on, a clerk like Einstein would simply be following "the inventor's way of thinking, and that will prejudice you. You have to remain critically vigilant." In this respect, Einstein's position at the patent office exploited a rather particular skill of his: the ability to size up a physical system at a glance and seize its essence—to understand how it was supposed to operate, and whether it would.

In part, it was this talent that Einstein brought to the image of a man falling. Although it existed only in his mind, Einstein understood at once that the image *worked*. In the equivalence between inertia, the tendency of an object at rest on a roof to remain there, and gravity, the tendency of an object not at rest on a roof to plummet toward the ground—between not falling and falling—he sensed he was glimpsing a likely answer to a certain question that had first arisen two years earlier, in the paper he'd now agreed to summarize for the *Yearbook*. Back then, Einstein had shown a mathematical relationship between a body at rest and a body traveling at a uniform—or nonchanging—velocity. Now he planned to do to those 1905 equations what he'd wanted to do from the start: extend them to include a body at rest and a body traveling at a *non*uniform velocity—that is, accelerating, which is what a man in the grip of gravity does. Before mailing off his *Yearbook* article in early December 1907, and emboldened by his vision of a man and a roof, Einstein added a section expressing his hope to be able to do just that, as soon as he got around to mastering the math.

The math: that element in any physics proof that was unquestionably necessary, but also neglectable. Einstein, at any rate, had neglected it. Even during his student years, Einstein had often skipped the lectures on mathematics, relying instead on the meticulous notes of a good friend. "I really could have gotten a sound mathematical education," Einstein later reflected. "However, I worked most of the time in the physical laboratory, fascinated by the direct contact with experience." For Einstein, mathematics, however intricate and noble and essential, had become a means to far more fascinating ends.

Once, though, it had been an end in itself. As a child, Einstein had regarded mathematics with a respect bordering on reverence, an extension of the feeling that had first stirred in him on that memorable day when his father had produced a compass for his amusement and young Albert had asked himself what could be determining the direction of the needle. Nothing, as far as he could see—and as far as he could see was therefore not far enough.

This suspicion was hardly original, and it might be only a slight exaggeration to say that at that moment the young Einstein joined the ranks of philosophers dating back at least to Plato, whose parable of the cave had provided an enduring metaphor for civilization's potentially futile but unavoidable dependence on sense perceptions in the search for reality: chained prisoners watching shadows on the wall and trying, from the evidence of those few flickers, to discern the shape of the hidden forms casting them, if not to discover the even more remote source of light. So what was the alternative to flicker divination or compass contemplation? The answer that Einstein would one day reach was just as unoriginal as his insight into the limitations of sense evidence—and just as eternal: mathematics.

At the age of twelve, Einstein picked up a geometry text, and it might be only a slight exaggeration to say that at that moment he joined a venerable mathematical tradition, too, though of more modern vintage than the ancient philosophical one. True, Plato had suggested to the members of his Academy that they seek mathematical patterns in the movements of the heavens—that they find out whether geometry, a word meaning "measure of the earth," might also apply to the sky. And they'd succeeded, somewhat. The correlation between the diagrams they could draw on paper and the motions they could see in the heavens was good enough to guide astronomers for well more than a thousand years in predicting eclipses and seasons, in scheduling harvests and holidays.

Even so, this correlation was of necessity only an approximation and an admittedly errant one at that. In the absence of knowing how the heavens actually work, the best that later mathematicians could do was, in their own terminology, to "save the appearances"—to describe spheres within spheres along which the sun, moon, and planets could travel, then to ascribe to these travels circular motions within circular motions, until the mathematical picture on paper began to bear some resemblance to what appeared to be happening in the heavens. As for what actually was happening, that was unknown, and, given its inaccessibility to investigation, unknowable.

But it *was* knowable—or as knowable as anything—and it was the modern scientific tradition embodying this belief in which the twelve-year-old Einstein enlisted. Before his introduction to the geometry text, the young Einstein had tried to satisfy his yearning for a secret order to things by cultivating a "deep religiosity," though his parents, Jewish by birth, were both "deeply irreligious." Bavarian law required that children receive religious instruction; Albert received his from a distant relative at the Einstein family home in Munich beginning around the age of seven. For years Albert followed religious prescriptions down to the least detail, refusing, for instance, to eat pork. This interlude ended when he began reading popular science texts and, as he later wrote, he "soon reached the conviction that much in the stories of the Bible could not be true." A six-day Creation, for instance, didn't match the way the world works, at least absent a leap of faith. Geometry, however, did—no leap of faith required.

Or rather, the faith that mathematics required was of a different sort. Forget friends. Forget playing outdoors. Only equations concerned Albert now. When he reached the age of thirteen and the time had come for him to begin the formal study of algebra and geometry at *Gymnasium,* he acquired all the textbooks on the syllabus in advance and then, over the course of one school vacation, methodically worked his way through every theorem, mastering each in turn on his own. Days at a time he spent alone, focusing on a single problem until he had arrived at a proof—a proof, moreover, that often differed from the one in the text but was, he saw, no less legitimate. Even the Pythagorean theorem he managed to prove in an original manner. Maybe what mattered in mathematics wasn't so much *how* one arrived at an answer but that there was an answer to arrive *at*: an unvarying relationship between the square of the hypotenuse and the sum of the squares of the other two sides, an equality that held for all right triangles, no matter what.

"The objects with which geometry deals," Einstein later wrote, describing this first encounter with the Pythagorean theorem, "seemed to be of no different type than the objects of sensory perception." In which case, was there a difference? Since the ancient Greeks had based the abstractions of geometry on the particulars of everyday life, why shouldn't the manipulations of geometrical shapes provide some insight into the workings of the universe? Were the secrets of the cosmos in fact accessible through math and therefore ultimately knowable?

This was the faith that mathematics in the modern era required: "faith in the existence of natural law." Einstein would one day write these words in honor of Johannes Kepler, the German mathematician and astronomer who, in the early years of the seventeenth century, became the first modern thinker to embrace this belief. What made Kepler's conviction especially remarkable, especially admirable, Einstein went on, was that he acted upon it even though he "lived in an age in which the reign of law in nature was as yet by no means certain."

All that mathematicians had proven before Kepler was that through geometry they could save the appearances—they could approximate the motions of the heavens. But as Saint Thomas Aquinas wrote in the thirteenth century, "The assumptions made by the astronomers are not necessarily true. Although these hypotheses seem to be in agreement with the observed phenomena, . . . perhaps one could explain the observed motion of celestial bodies in a different way which has not been discovered up to this time." By Kepler's day, however, those mathematical hypotheses of old didn't even have the advantage of being in agreement with the observed phenomena: The arrival of the seasons, over the course of centuries, had slowly diverged from the mathematical predictions embedded in the calendar. When Nicolaus Copernicus proposed a new system of the universe with the sun at the center and the Earth revolving around it, in his 1543 masterpiece of mathematical astronomy *On the Revolution of Celestial Orbs,* his primary motivation was still to "save the appearances": to find equations to account for what only *appears* to us to happen in the sky.

By the turn of the seventeenth century, however, Kepler was proceeding from a different assumption altogether. He believed not that the heavens merely appear to follow the laws of geometry, but that they actually do—and that he, therefore, could use geometry to describe the way the heavens not just apparently, but actually, work.

In these efforts Kepler enjoyed a tremendous advantage over all previous mathematicians: observational precision without precedent. In 1600 Kepler joined the staff of the Danish astronomer Tycho Brahe, who recently had become the imperial mathematician in Prague. Before that appointment, Tycho had spent more than twenty years compiling tables of planetary motions, courtesy of an immense endowment from the king of Denmark that had provided for the purchase of an island, the establishment of an observatory there to Tycho's specifications (complete with four separate observation rooms), and the construction, in a machine shop on the grounds, of the most elaborate and most accurate astronomical instruments in history.

What Johannes Kepler brought to Tycho's data, however, was something no less valuable: a mathematical precision equally without precedent. Like Copernicus, Kepler realized that a sun-centered universe would help his mathematical reconstructions of the celestial vault. But instead of ignoring any remaining discrepancies between his math and the data, Kepler took the radical step of attempting to reconcile them until they disappeared. If the heavens did indeed operate according to laws and if math could indeed describe those laws, he would just have to work the math, adjust it, fix it, until what his mentor Tycho had observed and what he himself could calculate matched *precisely.*

Tycho died in 1601 but not before issuing to Kepler the challenge of rendering the orbit of Mars mathematically. Kepler soon determined that two of Tycho's painstaking observations of the planet's position diverged from his own calculations by only 8 minutes of arc—out of 21,600 arc minutes in an entire circular orbit, and this at a time when 8 arc minutes represented a level of precision no previous astronomical observer had even approached, let alone applied as a standard of proof. Kepler nonetheless accepted the discrepancy as legitimate and returned to the calculations. His standard of math demanded as much.

Not that Kepler himself was thoroughly modern. Year after year, as he bent over the calculations that ushered in a new era of celestial mechanics, he kept one ear cocked for the music of the spheres—a mystical outward manifestation of the internal order of the universe that the ancients had hoped to hear. Even his inkling that the heavens should ultimately follow the rules of mathematics arose in part out of a conviction that the Creator had nested the orbits of His six planets, one within the other, using only the five regular geometrical solids—pyramid, cube, octahedron, dodecahedron, and icosahedron. And Kepler's allegiance to old prejudices wound up costing him dearly. At one point in his struggle to determine a mathematical formulation for the orbit of Mars, he arrived at what turned out to be the correct shape, but as it wasn't a circle, the symbol of celestial perfection since ancient times, he abandoned it. When, seven years and thousands of manuscript pages later, he arrived at it again, he recognized it as the ellipse he'd previously discarded. "Ah, what a foolish bird I had been," he noted.

It was in his exactitude, however, that Kepler set a new and wholly modern standard both for mathematicians and for mathematics itself—for what role mathematics could play in the investigation of nature. Ancient wisdom held that the movements of the heavenly bodies must be circular; once he had solved the Mars challenge, Kepler determined that they must all be elliptical. Ancient wisdom held that the motions must be uniform; Kepler held that they must be varying—that in the course of its orbit a planet slows down as it moves away from the sun and speeds up as it nears the sun: specifically, that a line between a planet and the sun sweeps out equal areas in equal times. Kepler published these first two laws in 1609. Ten years later he added a third: that the farther a planet's average distance is from the sun, the more time its orbit would take, and the nearer the average distance, the less time the orbit would take: specifically, that the square of the time a planet takes to complete one orbit around the sun is proportional to the cube of the average distance of the planet from the sun. *Here* is what the heavens actually, not apparently, do, because *here* was the match between observations and calculations to prove it.

But *how* do the heavens do what they do? For nearly two thousand years, a single answer had sufficed: The sun, moon, and six planets move along solid spheres. Never mind that nobody could see these spheres; they had to be there, because the objects had to be moving along *something*. In 1588, however, Tycho Brahe declared he no longer believed that such spheres existed, for after calculating the trajectory of the comet of 1577 he had discovered that its path should have sent it crashing through them, shattering one sphere after another. So what alternative to this age-old interpretation did *Astronomia nova* (*New Astronomy*), as Kepler called his 1609 treatise, have to offer? After reading William Gilbert's 1600 work *De magnete, magnetisque corporibus, et de magno magnete tellure* (*On the Magnet, Magnetic Bodies, and the Great Magnet Earth*), in which the English physician offered the radical proposition that the Earth itself is a magnet, Kepler thought he had the answer: Maybe the sun, too, is a magnet, exerting a force that keeps the planets in their orbits. But then where was the math to prove *this* proposition?

This essentially was the question that Isaac Newton was asking himself in the apocryphal-sounding but apparently true story of how one day in 1665 he saw an apple fall in his garden. In his time, Newton enjoyed the same advantage over Kepler that Kepler had enjoyed over his own mathematical and astronomical predecessors: observational precision without precedent. Following the invention of the telescope in the early 1600s, astronomers beginning with Galileo had produced evidence that not only had helped validate Copernicus's sun-centered interpretation of the cosmos but had provided a wealth of fresh numerical data to plug into any formula that mathematicians might devise in their attempts to find out why Kepler's laws worked. Furthermore, in their consideration of the workings of the sun, moon, and planets—a category that had come to include Earth, now that it was no longer at the center of the cosmos—Newton and his fellow Englishmen could draw on Descartes's proposition that the same laws of Euclidean geometry apply equally to the increasingly indistinguishable celestial and terrestrial realms.

First, they supposed that any such attractive force, magnetic or otherwise, would be emanating from the center of an object's mass—for example, from the center of the Earth—and not from its surface. Second, they supposed that this attractive force would follow the same formula as the one for measuring the brightness of the sun or a planet—that it would fall off by the square of the distance (the inverse-square law, as it came to be called). To test these two ideas, Newton compared the acceleration of an apple at the surface of the Earth—which is, after all, just another way of saying at a distance of one Earth radius from the center of the Earth—with the acceleration of the moon—which, according to the measurements of the day, he knew to be at a distance of about 60 Earth radii. Did these two examples follow the inverse-square law? Was the Earth's attractive influence on the moon actually $1/60^2$—or 1/3600—the Earth's attractive influence on the apple? Did the mathematical verification actually match the existing observations? Newton ran the numbers and reached his conclusion: "pretty nearly."

And now for the hard part: What shape *precisely* would a planet describe in space if it followed these two principles in its orbital relationship to the sun? This was the question the English astronomer Edmund Halley casually put to Newton in August 1684 during a visit to Cambridge. Newton just as casually replied that although he couldn't put his hands on the paperwork at the moment, he'd figured out the answer some years earlier: an ellipse. A dumbfounded Halley urged Newton to find the paperwork and publish his computations, which, if correct, would have to be nothing less than the mathematical confirmation of Kepler's laws (as Newton must have realized, though, characteristically, he didn't much care about the public reception of his work). Three years later and thanks to Halley's underwriting of the publishing costs, Newton produced *Philosophiae naturalis principia mathematica* (*Mathematical Principles of Natural Philosophy*). As Einstein once noted, the *Principia*'s law of universal gravitation allowed physicists for the first time in history to address not only the question of what the heavens actually do but also—and the emphasis is Einstein's—"the question *how the state of motion of a system gives rise to that which immediately follows it in time.*"

The faith that mathematics inspired was beginning to change. At the turn of the seventeenth century, Kepler had assumed that nature obeyed laws, and in so doing he shifted the study of celestial motions from an ancient emphasis on accounting for apparent *irregularities*—on saving the appearances—to a wholly new emphasis on capturing *regularities*—on applying three simple laws that matched actual observations. Now Newton had once again shifted the emphasis regarding celestial motions, from their regularity to their *predictability.* The faith that math now engendered in its practitioners was not only that they could account for whatever might be happening in the heavens but that they would even be able, with the right amount of information and the appropriate application of formulas, to anticipate what might happen *next.*

"An intelligence knowing, at a given instance of time, all forces acting in nature, as well as the momentary position of all things of which the universe consists, would be able to comprehend the motions of the largest bodies of the world," the French mathematician Pierre-Simon de Laplace wrote in the late 1700s. "Nothing would be uncertain, both past and future would be present." What's more, in both his 1796 popular treatise *Exposition du système du monde* (*The System of the World*) and his far more technical, five-volume *Traité de mécanique céleste* (*Celestial Mechanics*), which he published between 1799 and 1825, Laplace went a long way toward proving that such an ambitious program could be done. All that remained for succeeding generations of astronomers and mathematicians, as far as they were concerned, was to *do* it—to refine observations and calculations to an ever greater order of precision and then to clear up the remaining inequalities. Laplace himself, in 1785, showed that long mysterious variations in the orbits of Jupiter and Saturn were due to gravitational effects—that they weren't otherwise impenetrable irregularities but quantifiable, predictable, ho-hum regularities. Two years later he performed the same prodigious feat with the seeming irregularities of the moon. When two astronomers in 1846 independently used seeming irregularities in the orbit of Uranus to predict the existence and exact location of a *planet,* Neptune, the consensus was that Newton's universal law had achieved its greatest triumph. Irregularities were nothing but regularities waiting for an explanation. Two hundred years after Newton published the *Principia,* astronomers and mathematicians could content themselves that they had investigated just about everything there was to know about gravity, with one exception: what it was.

What was gravity? This essentially was the question that Einstein was asking himself when he "saw" a man fall from a roof. In his fantasy, Einstein was trying to figure out a way to incorporate gravity into the work on relativity he'd begun two years earlier. Just as someone on board the old Galilean ship would have as much right to think of the dock leaving the ship as the ship leaving the dock, so the man in free fall from the roof would have every right to think of himself as being at rest and the remainder of the universe as being in a state of motion. What would seem to someone else observing him (from the roof, say, or on the ground) like gravity would seem to the man like inertia—*and they'd both be right.*

This wasn't idle philosophizing on Einstein's part; this was sound physics, precisely because he had the math to prove it—or the beginnings of a math, anyway. When he imagined a man falling, Einstein also imagined other objects falling alongside him. The fact that two objects with different masses fall at the same rate in the absence of air resistance had been common knowledge for three centuries, ever since Galileo first put forth that proposition (though not after conducting the true-sounding but apparently apocryphal experiment of dropping balls from the Leaning Tower of Pisa). The implication that Einstein suddenly glimpsed from his chair in the patent office was that if the falling man saw other objects falling along with him—absolutely evenly, not diverging from his rate of acceleration in the least—they wouldn't seem to him to be falling at all. The quality of mass holding a man in place on a roof, or inertia, and the force propelling him downward, or gravity—the values corresponding to not falling and falling—weren't equal simply through coincidence, as physicists (including Newton, in the *Principia*) had supposed for the past three hundred years. They were two versions of the same thing.

After arriving at this insight, Einstein had to drop the subject for a few years while pursuing another question of physics and, not incidentally, advancing from the patent office in Bern to an associate professorship at the University of Zurich and then a full professorship at the German University in Prague. During this period Einstein also often found himself distracted by the demands that come with a growing professional reputation—lectures to give, scientific conferences to attend, job offers to consider. His marriage was badly fraying as well. But when Einstein returned to the problem of how to extend the concept of relativity, he did so in the same manner that characterized much of his professional life: like a man possessed.

He submitted a paper on the topic in June 1911, and he published two further exploratory articles on gravitation in February and March of the following year, and then, in a follow-up paper he wrote that July, confessed he was stumped: "I would ask all colleagues to apply themselves to this important problem!" Once, Einstein went mountain climbing with the French-Polish chemist Marie Curie, and seemingly oblivious to the rocks and crevasses as well as to his guest's difficulty in understanding his German, he spent much of the time talking about gravitation. "You understand," Einstein said to her, suddenly gripping her arm, "what I need to know is exactly what happens in a lift when it falls into emptiness."

In Einstein's imagination, the motions of a hypothetical elevator had replaced the earlier freefall. In effect, Einstein had taken that imaginary man out of the imaginary resistance-free air and locked him inside an imaginary laboratory, there to arrive (with Einstein's help) at real results. Upright in the elevator, seemingly at rest, the man would have no way of knowing whether he was experiencing gravitation or acceleration—whether the elevator was simply standing on the surface of the Earth or, assuming that it was moving at the proper rate (approximately 32 feet, or 9.8 meters, per second squared, the gravitational force at the surface of the Earth), rising through space. What the man would feel would be the same in either circumstance—a perfect illustration of the principle of equivalence that Einstein had intuited while sitting at his desk back in 1907.

For the sake of argument, Einstein imagined that the elevator *was* in fact rising—that it was hooked at the top to some giant crane raising it through space. Next Einstein imagined a beam of light piercing the moving elevator—entering through one wall, passing through the compartment, and exiting through the opposite wall. If the elevator were rising relative to the light source, Einstein concluded that the height from the floor at which the light entered on one side of the elevator would not be the same height at which it exited on the other side. That is, from the point of view of the passenger, the light would appear to bend.

Again for the sake of argument, Einstein imagined that the elevator was *not* in fact rising. He imagined that it was stationary on the surface of the Earth. And then he asked himself: Since the two hypothetical circumstances were supposedly equivalent—an elevator accelerating through space at the proper rate and an elevator standing on the surface of the Earth—wouldn't the same effect have to hold true for both? In other words, doesn't gravity bend light?

"Grossmann, you've got to help me or I'll go crazy," Einstein shouted one August evening in 1912, returning to Zurich for the first time in several years and bursting into the house of a longtime friend and former classmate. Marcel Grossmann and Einstein had been students together at the ETH (Eidgenössische Technische Hochschule, or the Federal Institute of Technology) in Zurich in the late 1890s, when it was called the Swiss Polytechnic. It was Grossmann who had lent Einstein his math lecture notes back then. When Einstein's contentious and irreverent behavior toward his professors cost him any chance of earning a recommendation for future employment, it was Grossmann who arranged for his friend to get a job at the Bern patent office. Now, a decade later, Grossmann intervened on Einstein's behalf again, securing for him a professorship at their alma mater. Grossmann—one of the founders of the Swiss Mathematical Society, a full professor of mathematics himself at the ETH, and, as of the previous year, the dean of the school's mathematics-physics section—was a pure mathematician.

Einstein explained to his friend his insight that gravity was nothing more than a useful fiction, a concept to help us make sense of a movement across space and over time that our poor three-dimensional sensory perceptions alone can't. Maybe if you could trace the trajectory of a man falling from a roof in four dimensions, it would just be geometry. But what geometry?

Back in 1901, Einstein, the young snob about mathematics, had written to his fiancée that their friend Grossmann would be getting his doctorate in non-Euclidean geometry. Then Einstein had added this confession: "I don't know exactly what it is."

He was now about to find out. Einstein readily conceded that the mathematical strategy he had provisionally adopted in trying to reconcile relativity with gravity was "not obviously permissible." The problem, Grossmann explained to him, lay in the limitations of the laws of Euclidean geometry. The geometry of Euclid was adequate for describing three-dimensional physical relationships on a two-dimensional surface—a canvas on an easel, for example. But what about a three-dimensional surface that curves inward on itself, such as a globe, or one that curves outward, such as a saddle? On the first "canvas," so-called parallel lines would meet, on the second they would diverge, and in neither case would Einstein's holy Pythagorean theorem hold.

"Compared to this problem, the original relativity theory is child's play," Einstein complained in a letter to a friend. Perhaps he'd misjudged mathematics. Perhaps he should have attended more of those lectures back at the ETH, when among the courses he was supposedly taking was one on, yes, non-Euclidean geometry. In the early nineteenth century several mathematicians had independently begun to explore a geometry that did *not* correspond to the objects available through sensory perception, and by the middle of the century the German mathematician Bernhard Riemann had succeeded in unifying the various approaches into one grand overarching system. Just as two-dimensional Euclidean geometry could approximate the three-dimensional world of matter in space, so three-dimensional Riemannian geometry could approximate a four-dimensional world of bodies moving through space, which is to say, over time. "Never before in my life have I troubled myself over anything so much," Einstein wrote in the same letter, "and I have gained enormous respect for mathematics, whose more subtle parts I considered until now, in my ignorance, as pure luxury!"

In 1905, when he'd tried to figure out the relationship between a body at rest and a body moving at an unvarying, or uniform, rate, he'd found that the resulting math, while simple, had done funny things to time. Now, as he tried to figure out the relationship between a body at rest and a body moving at a varying, or nonuniform, rate, he found that the resulting math was doing funny things to space. If his calculations were correct, space seemed to sort of want to curve.

Bent light? Curved space? Did any observations in the world of sensory perceptions possibly correlate to the extreme cases in nature suggested by this manner of math?

In fact, Einstein thought, there were three. One observation would have been too difficult to perform with any assurance of accuracy, given the present state of technology. Another would have required a total eclipse of the sun, not exactly an everyday event but a test to which he'd alerted his fellow physicists in an early draft of the theory: "I now see that one of the most important consequences of my former treatment is capable of being tested experimentally. For it follows from the theory here to be brought forward, that rays of light, passing close to the sun, are deflected by its gravitational field, so that the angular distance between the sun and a fixed star appearing near to it is apparently increased by nearly a second of arc." To a Berlin astronomer who Einstein thought might be in a position to organize a scientific expedition to photograph a total eclipse of the sun, Einstein could be downright plaintive: "Nothing more can be done along theoretical lines. In this matter it is only you, the astronomers, that can next year perform a downright invaluable service to theoretical physics."

A third test of his new version of relativity, however, lay within Einstein's immediate grasp. It, too, concerned an extreme case in nature, but it was one that Einstein, like Kepler three centuries before him, could try to corroborate by matching his calculations with observations that astronomers had already made with painstaking precision: the orbit of Mercury.

Einstein had recognized this possibility for his theory right from the start. On Christmas Eve of 1907, only a couple of months after he'd envisioned a man in free fall, he'd written to a friend that he was working on a "law of gravitation" by which he hoped to "explain the hitherto unexplained" discrepancies in the motion of Mercury—an infamous problem of the day. In 1840, the director of the Paris Observatory suggested to an assistant, the same Urbain-Jean-Joseph Le Verrier who in a few years became one of the discoverers of Neptune, that he try to arrive at a theory for the motion of Mercury—"theory" in this case meaning the kind of match between observations and calculations that Kepler had first demonstrated. The difficulties inherent in observing a planet so close to the sun had always complicated the construction of such a theory for Mercury, but by 1843 Le Verrier had nonetheless used what data he had available to him to publish a provisional version. When close observations of Mercury in 1848 failed to confirm his mathematical predictions, he returned to the problem, and eleven years later he finally published a comprehensive revision to his theory. It came, however, at a price: an irregularity he couldn't factor away.

Although the task had its roots in Kepler's faith in the regularity of nature's laws, the more immediate impetus was a typically post-Newtonian faith in the predictability of nature's phenomena. Through rigorous application of the inverse-square law, an astronomer should be able to account for the position of a planet at any given time, past or future. If the solar system consisted only of one planet and the sun—what astronomers call a two-body problem—the solution presumably would involve nothing more than plugging numbers for mass and distance into Newton's inverse-square formula. But as Laplace had persuasively demonstrated in his *Celestial Mechanics,* a clockwork universe is a mechanism so highly complex it borders on the inscrutable for all but the most arduous of mathematicians, and Le Verrier had to factor into Mercury's orbit around the sun its inverse-square gravitational interaction with every other planet as well as, by extension, their interactions with one another. What he found was that the point on its orbital path where Mercury comes closest to the sun—the perihelion—was changing over time, slowly but measurably, as a result of the combined gravitational effects of the other bodies in the solar system. By his calculations, Venus, though small of mass on the scale of solar system objects, accounted for more than half the total of this gravitational effect, or 280.6

seconds of arc, because of its proximity to Mercury. Jupiter, though distant from Mercury, accounted for just under a quarter of the total, or 152.6 seconds of arc, because of its enormous mass. The amount that other planets perturbed Mercury's orbit ranged from Uranus's 0.1 arc second to Earth's 83.6 arc seconds. Add them all up, and the total came to 526.7 seconds of arc per century.

At least, that was the number on paper. The number in the heavens, according to what Le Verrier had determined from hundreds of observations by astronomers at the Paris Observatory dating back to 1801 and by other astronomers dating back to 1661, was different: another 38 arc seconds more per century, later refined by the American astronomer Simon Newcomb to 42.95. Le Verrier tried adjusting the variables—an estimate of a planet's mass here, a margin of error in an observation there—but found that nothing was going to close that gap. Astronomers and mathematicians alike were stuck with an anomaly in the advance in the perihelion of Mercury.

It was a small anomaly, but, as Kepler had demonstrated and as several generations of natural philosophers had learned, irregularities are merely regularities in need of an explanation. And so astronomers looked for one. Perhaps previously unseen planetary matter was the culprit—a planet between Mercury and the sun (which Le Verrier in advance christened Vulcan) or a belt of asteroids near the sun or a moon of Mercury. For more than half a century, astronomers strained to find just such a "material cause," as Le Verrier called it; a few even claimed to have seen the mystery planet, but follow-up observations were never able to confirm these sightings. In the end, some astronomers reluctantly began to consider the one other, far less agreeable, recourse: an inadequacy in Newton's law of gravitation.

It was just such an inadequacy that Einstein thought he'd detected in his calculations. If space did indeed curve in the presence of mass, then by his estimation that curvature would be detectable in the rapid orbit of the comparatively minuscule planet Mercury, spinning deep within the great gravitational maw of the sun. And because astronomers had already performed the necessary observations of Mercury, all Einstein presumably had to do in using that planet to test his theory was to plug the numbers from the existing data into his new formula to see how well the math matched Newcomb's 43 arc seconds per century.

The result wasn't even close: 30 arc *minutes.* Einstein soon realized that he'd made a mistake and recalculated, and the answer this time—18 arc seconds—was indeed much closer to what he'd hoped. But it was no more satisfying, at least not in a universe operating on a Keplerian scale of precision.

So maybe Mercury wasn't a test. Maybe there was another, simpler explanation for the advance in the planet's perihelion, one that had nothing to do with Einstein's math. In any event, as Einstein continued revising his theory he no longer bothered to worry about Mercury. In September 1913, he proclaimed to a friend, "The gravitational affair has been clarified to my *complete satisfaction.*" Not quite, as it turned out. He continued refining, and the following year presented his results to a meeting of the Prussian Academy in Vienna, declaring the theory to be near to its final form. His fellow physicists weren't so sure; some thought they detected a mathematical error. Einstein didn't see it. A year later, he did. By then, however, it was too late: He'd already booked himself into four further addresses at the Prussian Academy.

On November 4, 1915, he appeared before the academy to confess that he had "completely lost confidence" in his theory. He was, however, working on a revision, which he broadly outlined. The following week, he returned to the academy to elaborate on the revision and to introduce some new equations. Later, at home, he made another adjustment, and then, recalling the earlier test, he sat down to see if just possibly his new version could withstand comparison with Mercury's motions. He plugged numbers in, ran the calculations, arrived at a result: 43 arc seconds per century, precisely.

Something burst inside him. At least, that's how he later described the feeling he experienced at that moment: a snap. His heart began beating erratically, as if the irregularity that he'd banished from the heavens with a few strokes of his pen had taken refuge in his rib cage. "For a few days," as he soon wrote a friend, "I was beside myself with joyous excitement." It had happened again—that experience of sensing that there was more to reality than meets the eye. But this time was different. He wasn't a boy of five or six confronting a compass for the first time or an almost adolescent contemplating an ancient theorem. This time he himself was the creator of scratches on paper that matched motions in the heavens. Out there were the shadows, moving across the celestial vault, and here were the forms casting them, these non-Euclidean geometrical constructs written in his own hand. And what was the source of light? Could it possibly be a mind that once had glimpsed a vision of a man falling?

So it might seem. One day a few years later, in the early autumn of 1919, Einstein was talking to a student in his office in Berlin, where he was now a professor of physics at the university, about a book that took exception to his general theory of relativity—as his 1915 effort had come to be called, to distinguish its account of matter in accelerated motion from the earlier, more specialized, or special, theory of 1905 that addressed only matter in uniform motion. Einstein and the student were in the middle of their discussion when Herr Professor casually reached for a telegram lying on a windowsill, handed it over, and said, "Here, perhaps this will interest you." The cable contained the results from an expedition that had attempted to observe the gravitational bending of starlight during a recent total eclipse of the sun—one of the three tests of general relativity that Einstein had identified from the outset. According to Einstein's general theory of relativity, the curvature of space caused by a massive giant such as the sun could be measured during a total eclipse by observing the displacement of the light from the stars near the sun's rim and comparing their positions with where they would be if the sun weren't there. The student saw at once that the eclipse measurements, performed by the English astronomer Arthur Eddington, closely approximated Einstein's prediction, and she began expressing her pleasure. "But I knew that the theory is correct," Einstein interrupted her. Oh? And what, she asked, if the observations had disagreed with his calculations? "Then I would have been sorry for the dear Lord," Einstein answered. "The theory *is* correct."

Theory alone, however, *wasn't* enough, and Einstein knew it. One reason he'd received this cable from Lorentz, in Leiden, offering news of a tentative confirmation of his theory was that, in a postwar era when the scientific community of Berlin still needed to rely on intermediaries for information from the outside world, he'd sent a letter asking for it. And despite the outward calm he affected before his student, the sentiment he exhibited in a note that he dispatched to his ailing mother during the same period probably more accurately reflected his response on receiving the cable: "Today joyous news."

He might have been wrong, after all. "I am now completely satisfied and no longer doubt the correctness of the whole system, no matter whether the solar eclipse observation succeeds or not": This was what he'd written to a friend in March 1914 about the general theory of relativity in its then current form; fortunately for Einstein, the outbreak of war in Europe that summer and unfavorable weather conspired to prevent two separate expeditions from making a single observation of the total eclipse in Russia on August 21. If either expedition *had* succeeded in taking a reading and if that measurement had been remotely accurate, Einstein's "theory" would have suffered a major setback, for it still predicted a deflection of sunlight half the amount that his final formulation called for, many torturous mathematical adjustments and more than a year later.

For that matter, Mercury alone wasn't enough, either. The match that Einstein had found between measurements of the advance in the perihelion of Mercury and his own calculations may have provided him, back in November 1915, with a satisfying, even transcendent, glimpse of the power of mathematics. In his 1916 article "The Foundation of the General Theory of Relativity," Einstein pointed out that his equations not only lead to Newton's law of attraction but go beyond it, and he explicitly mentioned their superiority in explaining the anomaly in the advance in the perihelion of Mercury. "These facts must, in my opinion," he wrote, "be taken as a convincing proof of the correctness of the theory."

Yet they were not so taken, and not just because the Great War was stifling the spread of scientific information. The discovery of Vulcan or some other kind of planetary material near Mercury, or the invention by another bright young mathematician somewhere of a principle just as radical in its own way as Einstein's emerging work, might still account for the discrepancy between the observations of Mercury and the theory of Newton. As even a longtime champion of Einstein, his fellow German physicist Max von Laue, wrote about general relativity in 1917, "This agreement between two individual numbers . . . does not seem to be a sufficient reason, even though note-worthy, to change the whole physical conception of the world to the full extent as Einstein did in his theory."

Exactly. By the autumn of 1919, Einstein's conversion to a faith in the superiority of mathematics over observation may in fact have been as nearly complete as the anecdote involving his remark to his student about the eclipse cable seems to indicate. But he also understood that he nonetheless would need something definitive—or at least something less potentially ambiguous than the evidence involving Mercury—to persuade his fellow physicists, scientists in general, and perhaps even the world to join him in shrugging off a conception of the universe that for more than two centuries had answered every question we could think to put to it.

It came, as the cable predicted, that autumn. "REVOLUTION IN SCIENCE," read a headline in the November 7, 1919, edition of *The Times* of London. "New Theory of the Universe. Newtonian Ideas Overthrown." *The New York Times* at least equaled the hyperbolic tone of that coverage: "LIGHTS ALL ASKEW IN THE HEAVENS—Men of Science More or Less Agog Over Results of Eclipse Observations." The results of the two British expeditions to observe the total solar eclipse on May 29, 1919, were in, and by all accounts they validated Einstein's prediction of how much the gravity of the sun would bend the light of stars. Einstein's theory, Astronomer Royal Frank Dyson proclaimed, was therefore "one of the most momentous, if not the most momentous, pronouncements of human thought."

And this was just the beginning. The announcement of the findings before a special joint meeting of the Royal Society and the Royal Astronomical Society on November 6, 1919, provided the history of science with one of its few indisputable before-and-after moments: "EINSTEIN V. NEWTON," as *The Times* of London summarized the situation in a headline a few days later. For months to come newspapers and magazines around the world ran articles trying to detail what Einstein's theories meant for science, for the man in the street, and for Einstein himself, who finally resorted to selling photographs of himself (and donating the proceeds to starving children). As *Scientific American* declared, "This report affords one of the rare instances in which a topic of pure science assumes sufficient journalistic importance to justify cable despatches a column long; and in this case at least the real importance of the discovery is commensurate with the popular attention which is devoted to it."

Even so, it *was* an extraordinary response—to such an extent that the frenzy itself soon became a legitimate news story. "This world is a curious madhouse," Einstein himself wrote to one friend in 1920, and to another, "I feel like a graven image." Some attributed the impact of the November 6 announcement to the isolating effects the war seemed to have had on the men of science who were now "more or less agog." Some found in a British affirmation of a German theory a powerful symbol of postwar rapprochement, while others found in the seeming absurdities of relativity an equally powerful symbol of postwar moral and societal disintegration.

But among physicists who were beginning to learn to monitor Einstein's intellectual development, the response was due at least in part to the mathematics itself and what *it* symbolized.

The faith that mathematics could inspire among its most ardent practitioners was beginning to change again. First Kepler had shifted the emphasis regarding a mathematical consideration of the heavens from the irregularity of the celestial motions to their regularity. Then Newton had shifted the emphasis from their regularity to their predictability. Now Einstein was shifting the emphasis further still, from the predictability of the motion of a celestial object to a prediction of its very existence.

Of course, this had been true—had been the triumph—of the discovery of Neptune some seventy years earlier: Calculations using Newton's laws had pinpointed the position of a planet that nobody had known was there. In fact, the president of the Royal Society made the comparison explicit when he called the observation of the deflection of sunlight during the May 29 eclipse, as *The Times* of London paraphrased his description, "the most remarkable scientific event since the discovery of the predicted existence of the planet Neptune."

Still, Neptune was a planet. Planets, people knew. Not so a curvature in space that could account for Mercury's eccentric behavior. That nobody could have predicted, except for one Albert Einstein.

The new faith that the mathematics of Einstein was beginning to inspire arose not from its ability to explain the irregularities in the motions of the heavens by predicting the existence of heretofore unseen yet familiar objects: a planet; asteroids; a moon. It arose instead from its ability to explain the irregularities in the motions of the heavens by predicting the existence of heretofore unseen yet *un*familiar objects—unknown, and previously pretty much unimaginable, phenomena: bent light; curved space.

This was what had so moved Einstein himself on that November evening in 1915, when his heart skipped many beats. He had explained an irregularity in the heavens by invoking a phenomenon the existence of which nobody else had ever predicted. But a further implication of this explanation had eluded him.

For the past two to three centuries, mathematicians had been proceeding under this Keplerian assumption: If it exists astronomically in the heavens, it can be captured mathematically on paper. Now, mathematicians explicitly following the lead of Einstein were beginning to wonder if they might proceed under a new assumption: If it can be captured mathematically on paper, then it exists astronomically in the heavens.

As is often the case with a bold new doctrine, some converts wound up taking it more seriously than the prophet who introduced it. In 1917, a year after publishing his theory of general relativity, Einstein wrote a paper addressing the theory's "Cosmological Considerations." Could he apply his ideas of space and gravitation to the way the universe works on the largest scale? Just as Newton had before him, Einstein realized that a universe where every object attracts every other object through the force of gravitation must sooner or later collapse in on itself. Einstein's solution was to introduce a mathematical term—a cosmological constant: "lambda, at present unknown"—that would serve to counteract this collapse, at least on paper.

But on paper, to some minds, was beginning to assume a reality all its own, especially after the November 1919 announcement of the eclipse results seemed to affirm the existence of bent light and curved space. One such mind belonged to Aleksandr Aleksandrovich Friedmann, a meteorologist as well as a teacher of physics and mathematics at the university in Petrograd. When he looked at Einstein's paper on the cosmological applications of general relativity, he understood why Einstein would want to introduce a constant to counteract the effects of gravity, but he also had to wonder what would happen if he removed this more or less arbitrary element. Friedmann did so, and what emerged in 1922 was a universe in motion over time: a universe that contracts perhaps but also possibly expands or even does both in periodic fluctuations. It was wild, completely, but after 1919 all cosmological bets were off. So, Friedmann had to ask: Could this universe on paper be the one that matched the actual motions in the heavens?

Einstein objected. First he found fault with Friedmann's math, publishing a terse statement to that effect, though the calculational error, alas, turned out to be his own. Einstein acknowledged as much a year later but privately continued to grumble that all Friedmann had done was perform a mathematical exercise without practical application. Yet even as Einstein was expressing these reservations, the American astronomer Edwin Hubble was determining that at least one of the smudges at the farthest reaches of the most powerful telescope in existence—the 100-inch-diameter mirror at the Mount Wilson Observatory in the mountains northeast of Los Angeles—is exterior to our own Milky Way system of stars. And if that one smudge, that one "island universe" of stars, is exterior to our own, then are at least some of the other smudges? They are, Hubble soon found, and then in 1929 he found something else—that those island universes seem to be receding from our own at a rate proportionate to their distance. In other words, the universe—this collection of smudges, each of approximately equal size and magnitude to our own smudge of several tens of billions of stars, including our sun—seems to be expanding.

Ah, what a foolish bird Einstein had been. Like Kepler three hundred years earlier, Einstein's allegiances to old prejudices wound up costing him dearly. Like Poincaré within his own lifetime, Einstein hadn't yet absorbed the possibilities of a new, emerging conception of the universe. Although he himself acknowledged that the insertion of lambda into his cosmological equations was "gravely detrimental to the formal beauty of the theory," he'd felt the need to introduce it "for the purpose of making possible a quasi-static distribution of matter, as required by the fact of the small velocities of the stars." The velocities of the stars were in fact small, relative to the motions of Earth. What wasn't small, however, were the velocities of the systems of stars that lay *beyond* our own stars.

Even as he was adjusting Newton's previously impregnable equations—and despite his professed faith in his own sacred math, which was pointing him in what proved to be the right direction—Einstein couldn't quite accept a conception of the universe that so radically contravened the existing evidence. He couldn't believe that the universe operated on a temporal scale thousands of times longer and a spatial scale billions of times larger than what astronomers had at that point observed.

The hold of the Newtonian conception of the universe on the imagination was simply too powerful. By now, the illusion of a clockwork cosmos whose pieces operated with exquisite regularity but whose whole didn't change over time was, as Einstein might have said, unrecognizably anchored in his unconscious.

In 1931 Einstein visited Hubble at Mount Wilson and considered for himself the empirical evidence, "which," as he immediately wrote in a paper, "the general theory of relativity can account for in an unforced way"—that is, without the introduction of the cosmological constant he himself had, well, forcibly introduced. His insertion of that constant, he now conceded, was the "greatest blunder" of his career. It had blinded him to an accomplishment of his that was arguably as impressive as the creation of a new theory of gravity. It was the creation of a new theory of the cosmos on the whole, which is to say a new cosmology, which is to say: a new science.

FIVE

THE DESCENT OF A MAN

In his mind, he was falling. Night after night; from a great height; falling. His waking hours were no less nightmarish. A few weeks earlier a woman had accused him of theft, and since then his eyesight had grown weak, and sometimes he saw everything as shades of gray. During this period his memory also had become unreliable. His ears rang, his tongue stuck, his stomach ached. His left hand and left foot had developed a tremor, and he found that, at the age of twenty-nine and after a ten-year apprenticeship as an engraver, he could no longer work. The entire left half of his body, in fact, felt as if he'd had a stroke. He experienced pains in the left side of his throat, in his left groin, and, if he walked for some time, in his left knee and left sole, yet if you were to pinch a fold of skin from anywhere on the left side of his body and pass a pointed needle through it, he would feel . . . nothing.

He was perfect. The examining physician had been looking for a human subject to present before the Viennese Kaiserliche

Gesellschaft der Ärzte (Imperial Society of Physicians), and for his purposes August P. was in several ways an ideal specimen. At a society meeting a week earlier, this physician—Sigmund Freud, then an accomplished thirty-year-old lecturer at the Vienna General Hospital—had summarized his findings from a recent research trip to Paris. Following his presentation, one of the elders of the Society had suggested that Freud return with someone whose symptoms resembled those of the Parisian patient under discussion—a man who actually *had* suffered a fall and who claimed to have lost the use of an arm as a result but who showed no physical evidence why that might be so.

For Freud, these appearances before the society in October 1886 had come at an opportune moment in his career; that April, not long after his return to Vienna from Paris, he'd opened his own private medical practice. Yet however potentially advantageous the timing might have seemed for a young professional to address the members of the society, one of the most prestigious medical organizations in Europe, Freud would have known not to request an appearance unless he had something new to report, and ever since the trip to Paris he'd been petitioning the society for an opportunity to present what he'd learned during his residency at the Salpêtrière, the legendary Left Bank medical complex, and especially what he'd learned by working with the world-renowned Jean-Martin Charcot, the charismatic director of the Salpêtrière and the leading neurologist of the age.

He'd learned something in Paris; that much Freud knew. During his first appearance before the Vienna Society of Physicians, Freud reported that back in Paris he'd learned to distinguish between two types of hysteria, *grande hystérie* and *petite hystérie,* each of which could be identified by specific sets of symptoms. A society physician responded that in his practice he'd examined hysterics whose symptoms didn't fit either category. Freud said that in Paris he'd learned that hysteria—despite its derivation from the Greek for "uterus," due to the ancient belief that the illness arose from a wandering womb—was not uncommon among men. Two members of the society responded that in their medical circles male hysteria was in fact a well-known condition, and one of them, a neurologist, pointed out that he'd published an article on male hysteria sixteen years earlier. Finally, Freud said that in Paris he'd learned that when treating some victims of trauma, such as the case he'd described—a man who had fallen from a scaffold—physicians might profitably conceive of the symptoms as forms of male hysteria. The chairman of the meeting, one of the four professors who had awarded Freud the grant money to travel to Paris, dismissed such speculation as premature at best. "I was unable to find anything new in the report of Dr. Freud," the chairman concluded, "because all that has been said has already long been known."

As was customary at society meetings, the questions and comments that greeted Freud were vigorous, contentious, adversarial. Even so, the reception wasn't remotely what he'd hoped. What had gone wrong? Even if he had miscalculated the specific noteworthiness of what he'd learned in Paris, he couldn't possibly have misunderstood the overall fact that it was significant. Perhaps by returning with a human subject such as poor August P. and demonstrating for his fellow society members some of the phenomena that he'd witnessed for himself in Paris, Freud hoped he might somehow communicate what he was struggling to put into words.

Ten years later, he was still struggling. It was this struggle that culminated, in 1895, in his attempt to write a "Psychology for Neurologists"—culminated, actually, in his abandonment of that project. When Freud decided to forgo any further explorations of the pathways of the brain in favor of the pathways of the mind, he was finally making a distinction that had been eluding him since his visit to Paris. It was a distinction, moreover, that forced him to reconsider virtually everything he'd thought he'd learned since then. Back in 1886, he'd managed to impress the Society of Physicians on his return visit with August P., if only because the sight of a man having his skin pinched, his earlobe twisted, his wrist punctured with a needle, his throat probed with a finger, and his nasal passage navigated by a small roll of paper, all in the absence of any apparent physical discomfort, couldn't help but make an impression. But Freud had presented no new information, nothing of substance to augment his presentation at the earlier meeting, no answer to the lingering question of the significance of what he'd learned in Paris—the question to which he would now have to return. For when Freud had arrived in Paris on October 13, 1885, he was planning to continue his already considerable and valuable research into the workings of the anatomical parts of the nervous system—neurology. By the time he departed Paris the following February 28, he'd diverted his attention to the study of the behavior generated by the nervous system—psychology. At a certain point during those four and a half months he had lifted his eye from the microscope and, for the first time in his adult life, left the laboratory.

Freud had first entered the laboratory twelve years earlier. "Nature!" he'd heard and was stirred. "We are surrounded by her, embraced by her—impossible to release ourselves from her and impossible to enter more deeply into her." It was upon hearing a lecturer read aloud this century-old essay in early 1873, as Freud later recalled, that his seventeen-year-old self had decided to enroll at the University of Vienna as a medical student. His goal, though, was not to "help suffering humanity," as he wrote to one friend at the time, because he'd never felt "any particular predilection for the career of a doctor." Rather, as he wrote to another friend, it was "to become a natural scientist"—to "examine the millennia-old documents of nature, perhaps personally eavesdrop on its eternal lawsuit, and share my winnings with everyone willing to learn."

It might be only a slight exaggeration to say that at that moment the young Freud joined a philosophical tradition dating back at least to Aristotle, who advocated learning about the natural world by investigating only what presents itself to our senses. Not for him the method of seeking deeper truths advocated by his teacher Plato, the discerning of the hidden forms casting the shadows on a cave wall. Chained as we are by our five senses, we will never experience these forms directly; therefore we can never know them with any certainty. Better, then, to concentrate our investigative resources on the shadows themselves and only on the shadows—the natural phenomena that we *can* experience directly through our senses, however inadequately. And where can one find such a concentration of investigative resources? Where does one go to concentrate one's own investigative resources? The answer that Freud discovered for himself was as predictable as it was profound: the laboratory.

A year after entering university, as he wrote to a friend a short while later, if "asked what my greatest wish might be, I would have answered: a laboratory and free time"—and it might be only a slight exaggeration to say that by this point he had joined a venerable physiological tradition, too, though of more recent vintage than the philosophical one. True, Aristotle had subjected to scientific examination the myriad animals and plants that he'd assembled in his garden for that purpose. But the particular type of dissection that Freud eventually found himself performing was human—a modern variation.

Before the sixteenth century, a lecturer on the human anatomy would stand before his students and read from an ancient text while a demonstrator pointed out the relevant anatomical details. In the late 1530s, however, a Belgian named Andreas Vesalius, then in his twenties and a professor at the University of Padua, elected to perform the dissections himself. The depth of information he discovered through this unorthodox manner inspired him to commit his findings to paper. The result, in 1543, was *De humani corporis fabrica* (*On the Structure of the Human Body*), the first modern comprehensive anatomy of man.

In this undertaking Vesalius enjoyed a substantial advantage over his predecessors and contemporaries: a pictorial expertise without parallel. Most if not all of the illustrations in the *Fabrica* were probably executed by the students of the incomparable Tiziano Vecellio—or Titian—in nearby Venice, working under the supervision and direction of Vesalius. The collaboration between art and science in the *Fabrica* led to not only an unprecedented degree of anatomical detail but a subtlety of interaction between anatomical parts that suggested not a cadaver in repose but a figure in motion—a living, if flayed, organism.

What Vesalius himself brought to the *Fabrica,* though, was something no less valuable: an observational method without parallel. By conducting dissections himself, he bridged not only the literal distance between the lecturer and the subject under investigation but a metaphorical distance as well—the one between contemporary investigations and ancient teachings. In the book's first edition, Vesalius cautiously repeated the age-old proposition that blood passes from one ventricle of the heart to the other through minute pores in the septum. In a second edition twelve years later, however, he wrote that "the septum of the heart is as thick, dense, and compact as the rest of the heart. I do not see, therefore, how even the smallest particle can be transferred from the right to the left ventricle through the septum." In this instance and numerous others, he believed that he could rely on the results of his own investigations, even when they contradicted ancient teachings or current wisdom, because, as he wrote, he'd "put his own hand to the business."

Not that Vesalius was altogether modern. In contemplating the question of what, exactly, endows a living object with life, he accepted the reigning explanation of "animal spirits" in the bloodstream—little creatures themselves endowed by the Creator with a life force. But his hands-on methodology nonetheless set a new standard for what role empirical evidence could play in the study of anatomy, if not in the investigation of nature altogether.

What was the human body? Lay it open on the examination table and see for yourself. What did it do? A century after Vesalius, Descartes described the workings of the heart and arteries—"which, as the first and most general motion observed in animals, will afford the means of readily determining what should be thought of all the rest"—but first he advised readers "who are not versed in anatomy, before they commence the perusal of these observations, to take the trouble of getting dissected in their presence the heart of some large animal." The motion that they'll find therein—the flow of blood into the heart, the flow of blood out of the heart—Descartes described in anatomical detail that left little doubt not only that it follows "from the very arrangement of the parts" but that he could render it with mathematical precision: "I could set out many further rules here for determining in detail when, how, and by how much the motion of each body can be changed and increased or decreased by colliding with others—in sum, rules which comprehend in a concise way all the effects of nature."

Even an answer to the question of *how* the body does what it does hadn't been long in coming. The English physician William Harvey, having studied in the same department of anatomy at the University of Padua where Vesalius had conducted his revolutionary investigations half a century earlier, provided it in 1628, in his *De motu cordis et sanguinis in animalibus* (*On the Motion of the Heart and Blood in Animals*): Blood circulates. Just as Vesalius had argued, it doesn't seep through the septum. Rather, Harvey concluded, it leaves the heart through the arteries and returns through the veins. Less than a decade after Harvey made this suggestion, Descartes cited it in arriving at his primarily mechanical conception of the human body: "The nerves of the machine which I am describing may very well be compared to the pipes of these waterworks; its muscles and its tendons to the other various engines and springs which seem to move them; its animal spirits to the water which impels them, of which the heart is the fountain; while the cavities of the brain are the central office."

Not that man is a mere automaton—an *entirely* mechanical creature. To Descartes, he clearly was not. Plot all matter on a graph according to three coordinates in space, grant it a cause-and-effect inevitability, and you still couldn't explain all that man is capable of doing. The brain of an animal might function as a clearinghouse: The outside world sends signals that activate the sense organs, which in turn send signals to the brain; the brain responds by sending signals that activate the motor organs, which in turn send a clear message from the inside world. In this manner the brain of an animal dictates when the creature eats, runs, blinks, excretes, sleeps, mates, stops functioning. The brain of man dictates in this manner, too. But: Man also has a capacity for reflection and reason—an inner life, a consciousness—and for Descartes this quality was sufficient to differentiate man from animal.

Descartes himself anticipated objections to this proposition. He invited them, he received them, and he responded to them, at least to his own apparent satisfaction. And then even he, twelve years after he first formulated the division between the *res extensa* and the *res cogitans*—between the extended thing and the thinking thing, between brain and mind—took the first step down the slippery slope and sought to locate the seat of the soul *somewhere.* He chose "the innermost part of the brain, which is a certain very small gland situated in the middle of the brain's substance and suspended above the passage through which the spirits in the brain's anterior cavitities communicate with those in its posterior cavities."

Over the following two centuries, few (if any) students of either the physical or psychical components of the human condition emulated this particular literal-mindedness, so to speak. Still, they honored the basic distinction between mind and brain that preserved man's exalted status among God's creatures. Although Goethe wasn't the author of the essay that had inspired Freud to enter the laboratory, as Freud erroneously believed, he did conduct research comparing the upper jawbones of humans and animals in the hopes of "bringing man and animals together, tracing them to be one." One, yes, but not the *same.* "Man is animal," Goethe also wrote, before immediately qualifying his claim: "an animal with a difference, singled out for higher things." Or as a contemporary of Goethe, the anthropologist Johann Gottfried von Herder, wrote: "At man the series stops." This, then, was Man at the turn of the nineteenth century: not separate, but not equal, either.

Not so when Charles Darwin got through with him. In 1859—when "I was already alive," as Freud liked to point out, though he was only three years old—Darwin published *On the Origin of Species.* Humans, in that work, received precisely one mention, a promise from the author that in future writings "light will be thrown on the origin of man and his history." In 1871, while Freud was in *Gymnasium,* those writings arrived: *The Descent of Man,* a two-volume work that Freud kept on his bookshelf until the end of his life. In this sequel of sorts, Darwin finally made explicit what he'd earlier left delicately, deliberately implicit—that the same process of natural selection he'd discovered in plants and animals would have to apply to *Homo sapiens* as well. "Man is not a being different from animals," was how Freud once summarized Darwin's message, before immediately *not* qualifying this claim: "or superior to them." Or, as Darwin himself once wrote of his own species, "I see no possible means of drawing the line and saying, here you must stop." No splitting the difference now. Here in the slop at the foot of the slippery slope was a new view of Man: not separate; equal: an automaton, even.

By the time Freud entered the University of Vienna in 1873—only fourteen years after the publication of *Origin* and only two after *Descent*—"the theories of Darwin, which were then of topical interest, strongly attracted me." Even if Freud hadn't yet discovered the theory of natural selection for himself, he soon would have, for Darwin's ideas were already driving the research programs that the impressionable young Sigmund encountered at the university. The same year that Freud arrived there so did Herr Professor Carl Claus, one of Darwin's most influential disciples on the Continent, who now became the director of the university's Institute of Zoology and Comparative Anatomy. Freud spent his first semester at the university fulfilling several basic medical requirements; then in the second semester he elected to take Claus's course, "Biology and Darwinism." In his fourth semester, the summer term of 1875, Freud switched from the medical program—or "zoology for medical students"—to zoology proper, a move that placed him directly under the supervision of Claus. The following year, Claus awarded Freud a research assignment at his department's new Marine Zoological Station in Trieste. Then Claus renewed the grant. And then—in March 1877, when Freud was not yet twenty-one—Claus presented to the Vienna Royal Academy of Science Freud's first scientific study, one that appeared the following month in the academy's *Bulletin* and one that placed Freud's work directly in the line of descent from Aristotle, who had himself unsuccessfully investigated the same subject: *"Beobachtungen über Gestaltung und feineren Bau der als Hoden beschriebenen Lappenorgane des Aals"* ("Observations on the Formation and More Delicate Structure of the Lobe-Shaped Organs of the Eel, Described as Testicles").

By this time Freud had once more modified his course of study at the university, leaving zoology for Ernst Brücke's Institute of Physiology—where the predisposition of the research assignments was, if anything, even more Darwinian. Thirty years earlier, in fact, Brücke was part of a group of young physiologists in Berlin who had anticipated (and, conceivably, influenced) Darwin's strictly physicalistic interpretation of man. "Brücke and I have sworn to make prevail the truth," wrote the promising neurophysiologist Emil Du Bois-Reymond in an 1842 letter, "that in the organism no other forces are in effect than the common physical and chemical ones." This vow came in direct response to the ancient "animal spirit" vagueness, to the "vitalism," that pervaded physiology even at this late date—what one prominent French physiologist could that same year still describe as "an imponderable agent known by different names such as *principle, agent or nervous fluid, nervous force, active principle of the nerves,* etc." Already Du Bois-Reymond (along with other researchers) had discovered that a nervous impulse initiates chemical changes in the nervous system. In 1850, another colleague, Hermann von Helmholtz, managed to measure the velocity of the nerve impulse in frogs: about 88 feet (or 27 meters) per second. So: They had matter, they had motion, and they had a club, the Berliner Physikalische Gesellschaft, dedicated to the proposition that they in effect would do for the physiology of man's inner universe what Laplace had done for the physics of the outer: to render it, once and forever, in purely mechanistic terms.

This was the foundation of the lectures in physiology that Brücke delivered and Freud attended at the University of Vienna, and this was the philosophy behind the research program that Brücke instituted and Freud now joined. To Freud fell the responsibility for determining whether the spinal cord of the lowly fish species *Ammocoetes petromyzon* shared certain features with the central nervous system of higher species, including, by extension, man. It did, he discovered. From there Freud graduated to a similar exercise on the nerve cells of the crayfish. When Freud eventually left Brücke's institute (and with it the university) for the Vienna General Hospital, in 1882, the next step in his researches have seemed as logical as only a generation or two earlier it might have seemed heretical, even nonsensical. "The subject which Brücke had proposed for my investigations had been the spinal cord of the primitive fishes," Freud later recalled, referring specifically to his transition from the university to the hospital, "and I now passed on to the human central nervous system."

The human central nervous system: Here, surely, was the seat of . . . something. Call it consciousness or the mind or the soul. Whatever it was, no longer would the ancients' description of "animal spirits" or Descartes's invocation of a mysterious *res cogitans* suffice. Whatever it was needn't be immaterial at all. Wouldn't be, in fact, if you followed the argument that had begun with Vesalius's examinations, found full three-dimensional expression with Descartes's descriptions, and reached new rhetorical heights and physiological depths in the pronouncements of the Berliner Physikalische Gesellschaft. What was consciousness? Never mind: What was *this*—this cell here, this fiber, this weblike structure, this lesion?

It was the answer to this question that Freud sought throughout the early 1880s and that he then traveled to Paris in the autumn of 1885 to continue seeking. What he found instead was the diminutive yet dominating figure of Jean-Martin Charcot, then performing at the height of his investigations into hysteria.

For thousands of years hysteria had served as what Freud once called "the *bête noire* of medicine"—as a catchall diagnosis that encompassed so many symptoms as to be virtually meaningless: "general nervousness, neurasthenia, many psychotic states and many neuroses." Not so at the Salpêtrière. There, hysteria had been "lifted out of the chaos of the neuroses" because there the great Charcot reigned, distinguishing historically indistinguishable ailments one from the next, christening them and categorizing them like some Napoleon of Neuroses, as he was commonly known, or, as Freud preferred to say, like Adam "when the Lord led him before the creatures of Paradise to be named and grouped."

What Freud had traveled to the Salpêtrière to witness was in fact a genesis of sorts: the birth of a new way of looking at the human brain. Like France itself, neurology had undergone a revolution only a hundred years earlier. Within the walls of the Salpêtrière—as visitors were reminded by a print that hung in the hall where Charcot lectured—the physician Philippe Pinel, acting in the same spirit as the crowds storming the Bastille directly across the Seine, had liberated the insane from their chains. No longer to be seen as the vessels of demons, the mentally unstable thereafter were treated as victims of anatomy, of imperfections in the brain—that is, as the province not of priests but of physicians. Upon his appointment to the Salpêtrière in 1862, Charcot transformed its several long seventeenth-century stone buildings from a vast poorhouse for four to five thousand elderly women—an incarnation more humane than the madhouse of old, perhaps, but hardly more scientific—

into one of the most influential research centers in the history of medicine. Charcot instituted case histories, autopsies, the admission of male patients; installed laboratories, consulting rooms, a photography studio, a museum, an outpatient service, an auditorium; and assembled an intellectually formidable staff that helped him establish what Freud could still accurately describe, in those years, as the "young" science of neurology.

Freud arrived in Paris hoping to pursue his own research in this field. Yet less than halfway through his residency there, as he made clear in letters to his fiancée Martha Bernays, he was ready to return home, for reasons that were at least partly personal. The city was so expensive and he was so poor that he resorted to buying a pen with a finer tip in order to save on paper and postage and to perching in the cheapest seats at the theater, "really shameful pigeon-hole loges." He reserved his strongest frustration, however, for the people of Paris, who "seem to me of a different species from ourselves." The shopkeepers "cheat one with a cool smiling shamelessness." Everybody, in fact, is duplicitous: "all you hear is 'charmé' (which is not true)." "As you realize," he wrote on December 3, seven weeks after his arrival in Paris, "my heart is German provincial and it hasn't accompanied me here; which raises the question whether I should not return to fetch it."

Yet however severe his homesickness, Freud found himself on his arrival in Paris confronting a professional crisis that was even more unsettling. For Freud, the problem wasn't simply that the laboratory facilities at the hospital were disorganized and "not at all adapted to the reception of an extraneous worker." In fact, in the first weeks of his stay he did manage to perform some important research on infantile brains, his area of specialization. The problem wasn't even that the celebrated

Charcot was too busy to bother with the likes of Freud, one of many foreign visitors who at any moment were making the pilgrimage to Paris, the capital of the scientific world, and to the Salpêtrière, the cathedral devoted to the pathologies of the nervous system. At least Freud had the opportunity to observe Charcot on a regular basis. "I sometimes come out of his lectures as from out of Nôtre Dame, with an entirely new idea about perfection," Freud wrote to his fiancée toward the end of November, right around the time that he was thinking of returning home. "But he exhausts me; when I come away from him I no longer have any desire to work at my own silly things." The problem was that, as Charcot liked to say and as Freud quoted him on his return to Vienna, "the work of anatomy was finished."

Freud interpreted this remark as an understandable use of hyperbole on Charcot's part—"as no more than an expression of the turn which his own activities have taken." Still, at the Salpêtrière at least, it reflected an unavoidable truth. Something new was happening here, and it wasn't happening in the labs where Freud had come to toil.

Each Monday at the Salpêtrière Charcot delivered a formal lecture, always to a standing-room-only crowd, often in the presence of the upper echelon of Paris society. On Tuesday he conducted what amounted to an informal lecture—an on-the-spot examination of a previously unseen case from the outpatient department that tested his formidable diagnostic powers. The rest of the week he saw patients in his office, accompanied by whatever members of the hospital staff wished to observe him at work, including, often, Freud. By all accounts Charcot used these various occasions to deliver something more than a masterly discourse on the several branches of medical knowledge that he himself had pioneered. Instead, he presented memorable spectacles that for sheer showmanship rivaled anything in a Parisian theater—and this time Freud had the equivalent of a front-row seat.

When Charcot needed to illustrate a specific stage or characteristic of hysteria, he could draw upon the vast population of human misery at the Salpêtrière to produce a patient exhibiting the precise corresponding symptom. He himself could expertly mimic symptoms. He dazzled his audiences with a recent invention, the photographic projection, and he devised ingenious demonstrations—for example, highlighting the tremors of various diseases by dressing several women patients in elaborate hats, then inviting the audience to observe the differences in how their feathers shook. Most dramatic of all, though, were the results he produced in hysterics who showed up unannounced at the door to the Salpêtrière, sometimes after traveling halfway around the world just to meet the famous French healer. Under Charcot's direction, the paralyzed regained the use of their limbs and rose from their stretchers. The lame threw aside their crutches and walked.

"Paris is simply one long confused dream," Freud wrote Martha on December 3, in the same letter in which he wondered whether he should return home, "and I shall be very glad to wake up." Wake up he did, that very week, though not by abandoning Paris. Overhearing Charcot complain about not having heard from his German translator in a long time, Freud offered his services, and from that moment forward his experience of Paris was transformed. Freud became a frequent guest at Charcot's palatial home on the Boulevard Saint-Germain. "You can probably imagine my apprehension mixed with curiosity and satisfaction," Freud wrote Martha in anticipation of his first such visit. "White tie and white gloves, even a fresh shirt, a careful brushing of my last remaining hair, and so on. A little cocaine, to untie my tongue." Three weeks later, he was sighing like a born boulevardier: "What a magic city this Paris is!"

However redeeming socially his contact with Charcot might have been, it turned out to be even more rewarding professionally. That first week of December 1885, Freud effectively passed from observer to participant—from spectator, albeit one privileged enough to occupy a front-row seat at the most spectacular medical drama then running, to player on the stage. Now, working at Charcot's side and often adding his own insights to the great man's diagnoses, Freud had the opportunity to observe and investigate for himself the bizarre nature of hysteria.

How to account for these phenomena? What could possibly be their physiological cause? Charcot couldn't say, and he professed not to care. In the laboratory, anatomists were trying to locate the inner disturbances—the lesions—corresponding to the kinds of behavior he encountered everywhere he turned at the Salpêtrière. So far, they had failed. Their craft, obviously, wasn't yet sufficient. That the outward manifestations of hysteria ultimately would be traceable to such lesions, once instrumentation had improved to the point it permitted such an acute examination of the neuroanatomical matter, Charcot had no doubt. Until then, though, Charcot contented himself with inspecting and distinguishing and cataloguing what he *could* examine: the symptoms themselves.

Charcot was right, Freud realized. Not that the work of neuroanatomy was done; no impartial observer would reasonably conclude that this particular science was at risk of exhausting itself. At that moment, after all, neuroanatomy verged on "the very threshold of mind." Rather, the news from the Salpêtrière was that Charcot was ushering out of the shadows of the madhouse and onto the stage of mainstream medicine a whole class of strange behavior defying physiological explanation and that if you wanted to begin to understand hysteria, you had better see it for yourself. It wasn't Paris that Freud needed to leave. It was the lab.

In a way, Freud had been leaving the lab for some time now. Four years earlier, at the age of twenty-six, he'd met Martha, the woman who quickly became his fiancée, and at once he'd begun planning for the day he would be able to support a wife and family. Brücke, by this time his mentor as well as the director of the laboratory where Freud had just spent what he later called the six happiest years of his life, counseled him for financial reasons to redirect his medical career from theory to practice—from conducting research in a laboratory to examining patients in a clinical setting.

Freud had never thought of himself as a healer per se. In choosing medicine as a career he had entered a profession and a tradition in which, as he knew, he need never see patients. Since approximately the midpoint of the century, this was an option that physiologists had been able to exercise. Brücke himself, coming of academic age in the 1840s, had never in his life seen a patient. Now, though, Freud followed Brücke's advice and applied to the General Hospital of Vienna, where he soon received the title of assistant physician. For the next three years he patrolled the wards, administered to patients, and made expert diagnoses.

Even then, though, Freud's love of the lab remained evident. At the hospital, in a setting where he might have acquired all the experience anyone could ever want in preparation for establishing a private practice, he soon gravitated to the Laboratory for Cerebral Anatomy. There he distinguished himself through his studies of the medulla oblongata, the patch of matter connecting the spinal cord and the cerebrum. He also made a significant scientific discovery there while researching the effects of cocaine—that the drug can serve as an anesthetic during surgery. (To his horror and enduring professional regret, Freud inadvertently made a gift of this key insight, and therefore the priority for the discovery, to a colleague.) Then, even as Freud was nearing the completion of his three-year transition from Brücke's laboratory to private practice, he traveled to the Salpêtrière to pursue not medicine in a clinical sense but research.

Even before his trip to Paris, Freud managed to fashion a kind of compromise with himself, one that his exposure to Charcot's expertise as a clinician could only have reinforced: He would work in the lab *and* open a private practice. Either way, he would concentrate his attentions on his specialty, the nervous system. Immediately upon his return from Paris in 1886, he assumed a title he'd previously accepted, as the director of the new neurological division at the Max Kassowitz Institute for Children's Diseases, where he would be able to continue his research on the human central nervous system. He also sent out notices informing the public that as of April 25, 1886, Dr. Freud of Vienna I, Rathausstrasse 7, was open for business.

On a practical level, the compromise worked. Within four months Freud's income from his private practice was both sufficient and stable enough that he was able, finally, to begin planning his wedding. On a more personal level, however, he found the experience of treating patients enormously frustrating.

In good faith he had turned to an 1882 textbook by Wilhelm Erb, *Elektrotherapie,* which, as he later recalled, "provided detailed instructions for the treatment of all the symptoms of nervous diseases. Unluckily," Freud went on, "I was soon driven to see that following these instructions was of no help whatever and that what I had taken for an epitome of exact observations was merely the construction of fantasy."

The terminology was intact—the form that the sciences had assumed over the preceding two to three centuries—but not the content. Freud, coming to the treatment of patients from a background in research that held the scientific method in highest esteem, was appalled. This wasn't science. This was guesswork in the guise of science.

He might have anticipated as much, given what he knew of the history of the Salpêtrière over the preceding century—how recently it was that Pinel and Charcot had liberated mental illnesses from the straitjacket of theological demonizing. And he might have anticipated as much, given what he must have experienced firsthand as an assistant physician at Vienna General Hospital in the early 1880s, at which point it enjoyed one of the finest reputations in the world among medical practitioners and patients alike . . . but only because a handful of physicians in the past several decades had deliberately mounted an effort to modernize it: to replace conventional wisdom with empirical evidence, magic with microscope. Even so, Freud had never before fully appreciated the *extent* of the ignorance pervading the healing profession on which he so recently had staked his future. As he later wrote about Erb's textbook, "The realization that the work of the greatest name in German neuropathology had no more relation to reality than some 'Egyptian' dreambook, such as is sold in cheap bookstores, was painful."

Freud found somewhat more success with several of the other standard procedures of the day for treating nervous disorders—hydrotherapy, massage, a rest cure. Still, by late 1887, otherwise unable to help his patients to the extent he wished yet needing more than ever to make a living (his first child, a daughter, had been born in October), Freud decided to try hypnosis.

Freud had seen hypnosis in action the previous year at the Salpêtrière, where Charcot used the technique to illustrate and validate the stages that he believed he had discerned in hysteria. As was the case with hysteria, the reclamation of hypnosis from the depths of scientific ignominy had come about primarily due to the influence of Charcot, who in 1882 had argued on behalf of its potential investigative value before the Académie des Sciences in Paris. The reaction to his presentation, as well as to the revival of hypnotism it incited, was by no means uniformly positive. Abuses both medical and moral surely waited, warned a vocal group of critics that included a former teacher of Freud's, Theodor Meynert, who fumed that under hypnosis "a human being is reduced to a creature without will or reason." Even one of the most ardent advocates of hypnotism admonished his followers that they risked making of a subject under hypnosis "a decapitated frog."

Well, yes. From the point of view of someone with Freud's background in dissection and human anatomy, that would have been precisely the advantage of this tool—the ability it afforded an investigator, as Freud later wrote, "to experiment on a human mind in a way that is normally possible only on an animal's body." In Paris, at the Salpêtrière, Freud had seen Charcot use hypnotism to illustrate and validate the stages he believed he'd discerned in hysteria. First Charcot would induce, one by one, the symptoms of hysteria in a patient, and then, one by one, he would remove those symptoms—that is, he used hypnotism in an almost diagnostic fashion. On his return from Paris in 1886, Freud had included a description of Charcot's use of the technique in the report he filed at the University of Vienna, recording his own "astonishment that here were occurrences plain before one's eyes, which it was quite impossible to doubt, but which were nevertheless strange enough not to be believed unless they were witnessed at first hand"; he'd twice lectured on the topic in Vienna shortly thereafter, once at the Physiological Club on May 11 and again before the Psychiatric Society on May 27; and by the end of the following year he'd contracted to translate and contribute a preface to the second edition of *De la suggestion et de ses applications à la thérapeutique,* by Hippolyte Bernheim, the director of a hospital in Nancy.

This last task Freud undertook only for the money, according to a letter to a friend. Bernheim believed that all hypnotic effects were purely the results of suggestion and that anyone therefore could be a subject. Charcot, however, believed—and Freud the physiologist had to agree—that hypnotic effects were all due to alterations in the conditions of the nervous system, though Freud wasn't so sure about Charcot's contention that they required a genetic predisposition on the part of the subject and that this genetic predisposition applied only to genuine hysterics. In the preface to his 1888 translation of Bernheim's *De la suggestion,* Freud stated that Charcot's and Bernheim's individual conceptions of hypnosis couldn't both be right, and then he diplomatically tried to outline the merits of each without disparaging Bernheim's too much. Even then, though, Freud was willing to at least consider splitting the difference: "We must agree with Bernheim, however, that the partitioning of hypnotic phenomena under the two headings of physiological and psychical leaves us with a most unsatisfied feeling: a connecting link between the two classes is urgently needed."

Whatever his initial reservations about translating Bernheim, the assignment soon rewarded Freud in ways he couldn't have imagined. What Bernheim practiced was something altogether different from Charcot. As Freud himself wrote in the preface, Bernheim's "use of hypnotic suggestion provides the physician with a powerful therapeutic method"—that is, one Freud could use on patients in his private practice.

Freud, of course, already had heard about the therapeutic potential of hypnosis six years earlier from his friend Breuer, who told him about the case history that Freud later called Anna O. But when Freud had mentioned the therapeutic potential of hypnotism to Charcot, he "showed no interest," as Freud later recalled. As a result, Freud had "allowed it to pass from my mind"—at least until the final weeks of 1887. Perhaps not coincidentally, it was at precisely this point in Freud's life—just as he was contracting to translate and write a preface to Bernheim's book and therefore presumably reading it—that Freud tried using hypnotism on his patients in the therapeutic, *curative,* manner that Bernheim, rather than Charcot, advocated.

Freud now proved to be only sporadically more adept at inducing somnambulism than he'd been at relieving his patients' sufferings through electrotherapy. In an effort to improve his hypnotic technique, Freud traveled back to France in the summer of 1889 and studied with Bernheim. "But as soon as I tried to practice this art on my own patients," Freud later recalled of the period following his return from Nancy, "I discovered that *my* powers at least were subject to severe limits, and that if somnambulism were not brought about in a patient at the first three attempts I had no means of inducing it." Still, the therapeutic results when he did manage to hypnotize a patient were positive enough and the overall potential for the technique encouraging enough that Freud continued to persevere, until the day he didn't.

It came in the autumn of 1892. The patient was Fräulein Elisabeth von R., a twenty-four-year-old complaining of pains in her legs and yet one more patient proving resistant to Freud's hypnotic ministrations. On this occasion, though, Freud remembered something he'd witnessed during his time in Nancy. It concerned the phenomena surrounding Bernheim's use of *post*hypnotic suggestion: not simply a command to raise your hand or lower your hand while under hypnosis; nor simply a command to raise or lower your hand half an hour from now, or tomorrow; not even what did indeed happen half an hour later, or the following day, when you raised or lowered your hand; but what came next: the denial of any knowledge of why you'd raised or lowered your hand—and then, under intensive questioning and after honestly continuing to deny any such knowledge and even in an entirely wakeful state, "lo and behold!" (as Freud put it): You'd give the explanation anyway.

And how had Bernheim overcome a patient's insistence that she was unable to remember the suggestion he'd given her? It was quite simple, really. He'd laid his hand upon her forehead and insisted she could. And then she did.

Now Freud tried the same trick. Leaning forward to place his hand on Fräulein Elisabeth's forehead, Freud said of the source of the trauma, "You will think of it under the pressure of my hand. At the moment at which I relax my pressure you will see something in front of you or something will come into your head. Catch hold of it. It will be what we are looking for."

And so it was. The memories came—maybe not at first but in time and then every time Freud repeated this variation of his usual technique. Then he tried holding the patient's head between his hands, and still the memories came. After a while, he stopped exerting pressure—physical and otherwise. He kept his hands to himself. Also, often, his thoughts. Instead, he sat back, out of sight, and let the patient do the talking.

And as he did so, he couldn't help noticing a similarity between the processes that must be behind a patient's defense against or repression of the initial trauma and the processes that must be behind the examples of posthypnotic suggestion that he'd witnessed during his stay with Bernheim.

Consider: If a subject could carry around a suggestion (to raise a hand, for instance, or say a phrase) for minutes, hours, or even days, until some external trigger brought it forth, then surely the facilitating alteration in the brain wasn't the spontaneous irruption of a lesion but something far more subtle—some change not in the structure of the brain but in *the way it works.* But what exactly does it mean to say "the way the brain works"? If not lesions, then what physiological change does the brain undergo in such circumstances?

This question returned Freud right back to where he started this investigation: Paris. Back in February 1886, during his final days at the Salpêtrière, Freud had mentioned to Charcot that he'd noticed a peculiarity regarding the nature of hysteria, and Charcot, to Freud's immense pleasure, not only encouraged him to write up his observation but promised to arrange for its publication. Freud's insight, courtesy of his knowledge of anatomy, was this: The way the symptoms of hysteria work didn't always match the way the structure of the human body works. An hysteric's loss of the use of an arm or a leg, for instance, corresponds to the common assumption that the limb extends as far as the shoulder or the hip, *not* to the actual distribution of nerves, which in fact extend beyond the shoulder or the hip.

Freud had begun writing the article after his return from Paris, and in subsequent years he continued to revise it. In an encyclopedia entry he wrote on "Hysteria" in 1888, he'd included the observation that the hysteric's paralysis of an arm or a leg, for instance, is often as accurate about "the structure of the nervous system as we ourselves before we have learnt it." Freud didn't finish the article he'd mentioned to Charcot until 1893, when he took the three sections he'd previously written and added a fourth.

It appeared in July 1893 in, as Charcot had promised, his journal *Archives de neurologie.* The following month, Charcot left Paris on holiday with a couple of colleagues, checked into a hotel room, and, the following night, at the age of sixty-eight, died. In the copy of the issue of *Archives* that he left behind in his library, however, Charcot had traced two heavy markings in the margin next to the passage regarding the seeming lack of correspondence between hysteria and anatomy, which comes at the conclusion of the third section of the article. Presumably he then proceeded to the brief fourth and final section, which Freud had recently added, and read this: "I will attempt to indicate, finally, what the lesion that is the cause of hysterical paralyses might be like. I do not say that I will show what it *is* like"—because he couldn't. Instead, Freud wrote, what's affecting the patient "will be an alteration of the *conception,* the *idea,* of the arm, for instance."

Psychical? Physiological? Phenomena of the mind? The brain? How could you tell the difference? *Was* there a difference? Fundamentally, no, if you followed the antivitalist line of reasoning—the physical and chemical chain of command from neuron to neuron. If not a lesion, then *some* physiological alteration somewhere in the circuitry of the nervous system must be taking place to create these psychical impressions. But what was changing in Freud's estimation was just how fundamental fundamental could be: the way the brain works, indeed.

In effect, Freud followed the advice that Du Bois-Reymond had offered back in 1842, in framing his and Brücke's antivitalism doctrine: If you can't find "the chemical physical forces inherent in matter, reducible to the force of attraction and repulsion"—if you couldn't be the Newton of neurology—then you should "assume new forces equal in dignity." If Freud couldn't quite trace the path of the "nervous energy" that he'd optimistically labeled "Q," for Quantity, in his "Psychology for Neurologists," then he would think of it as "psychic energy," and try to trace the path of *that*.

But for Freud, this wasn't merely a question of semantics, a matter of simple subsitution, one word in one conext for another in another. He would need to describe the workings of the mind in psychical language not just because he couldn't do so physiologically but because the workings of the mind *were* psychical. His patients were at the mercy not of lesions but of ideas. "Hysterics suffer from reminiscences," he wrote in his 1895 collaboration with Breuer, *Studies on Hysteria*, finally articulating the answer to a question that had first arisen ten years earlier. Where were August P.'s lesions? Maybe they were all in his mind.

No wonder *Studies on Hysteria* had contained an odd paradox, one that Freud had seen fit to address in print. In preparing the book it was Freud who pressed for a physiological description of psychic processes and it was Breuer who insisted otherwise. "In what follows," Breuer wrote in the opening paragraph of his section on theory, "little mention will be made of the brain and none whatever of molecules. Psychical processes will be dealt with in the language of psychology; and indeed, it cannot possibly be otherwise." Yet it was Breuer who wound up invoking the language of physics and chemistry on a regular basis, and it was Freud who wound up confessing that "it still strikes me myself as strange that the case histories I write should read like short stories and that, as one might say, they lack the serious stamp of science."

So what was a scientist to do? As he sat in his study, listening to his patients tell their stories, trying to encourage them to reach back farther into their memories for the origin of their symptoms—that inciting incident or idea which he hoped he might then help them rob of its influence over their lives—Freud began to notice how often they refused to cooperate. His patients promised to do their best, they said they understood the importance of reporting to him whatever might come to mind—"irrespectively of whether it seemed to them relevant or not, and of whether it was agreeable to them to say it or not"—and at his insistence they would echo their willingness to try. And then—

And then they changed the subject. They made excuses. They censored themselves. They talked about the noise from the clock in the next room, or about the difficulty of trying to think of something useful, or about nothing. They just lay there, silent. They *resisted.*

And *then.*

And then his patients went and said it anyway. After which what they said was this: "I couldn't believe it could be that." Or this: "I hoped it wouldn't be that."

They couldn't believe it could be, hoped it wouldn't be, *that:* the truth, even if it was still traveling incognito, "like an opera prince disguised as a beggar," or even if what came to mind wasn't the inciting incident or idea at all but merely one more step along the path back toward it. Still, Freud was beginning to understand that as long as a memory lay along that path, his patients would do anything to sidestep it, at least as long as the inconvenience, the discomfort, of avoiding it was less than the pain of confronting it. And then, when at last they could no longer avoid it, when they had recognized it and identified it for the distressing thought it was, they would say, "I could have told you that the first time." And then Freud, incredulous, would answer, "Why didn't you say it?"

Why *didn't* they say it? Because, clearly, they couldn't. Because something was preventing them. They knew what they needed to say; yet they didn't know. This "not knowing" suggested to Freud, writing in the *Studies on Hysteria,* a "not wanting to know"—"a not wanting," he went on, "which might be to a greater or less extent conscious." In this regard the conflict between his patients and himself reminded him of the one between the patients and themselves: The withholding of information from him wasn't necessarily deliberate, just as when the original incompatible incident had occurred or idea had arisen, the suppression of the same information from themselves hadn't necessarily been intentional.

And then, all at once, Freud knew why they didn't say what they knew the first time he asked for it: because whatever was preventing them from identifying this incident to him—the resistance—"must no doubt be the same psychical force" that first had exiled it from their own consciousness and, ever since, had kept it there, repressing it.

Not all patients resisted. Those he'd hypnotized, Freud found, would indeed under his command locate one pivotal memory after the next until eventually they reached the origins of their hysteria. But those he hadn't hypnotized—those who had proven difficult or impossible to hypnotize through either their own dispositions or his limitations as a hypnotist—did resist. When Freud had to resort to the "pressure" technique of placing his hands on the heads of his patients and instructing them to concentrate—that is, whenever he didn't have the overpowering force of hypnotic suggestion at his disposal—he found himself at the mercy of the resistance, just as his patients must have felt themselves at the mercy of the repression.

The pressure technique, then, seemed to have a distinct advantage over hypnotism. True, hypnotism offered him and the patient more direct, less inhibited access to the original inciting incident or idea. Freud could put the patient under a spell and the two of them could get where they needed to go, speeding past the inhibitions. But if Freud was right that the force of repression that kept these incompatible thoughts hidden from consciousness was manifesting itself in his office as resistance, then the pressure technique offered something that hypnosis could not: direct access to the inhibitions themselves.

For patients, the advantage of this slower, more arduous strategy was that it allowed them a deeper understanding of their torments, one that couldn't help but have a salutary therapeutic effect. But for Freud, the advantage of the pressure technique over hypnotism was nothing less than the chance to make a study of the mind in much the same way that he'd once dissected the brain.

This realization, as he recalled years later, represented for Freud "the triumph of my life." Prohibitions or commands on his part went only so far; the technique that Freud was refining now was as much a means of therapy as hypnosis had been, and, if he were right, it was something else, too—and something more: a means of investigation. By listening to his patients *not* do as they were instructed, by listening to them resist his suggestions or even their own hunches, he could observe for himself the defense, the repression—the very force that had rendered the trauma, however active, absent. A chair, a couch, and more than meets the ear—Sigmund Freud had found his way back to the lab, after all.

What was the human mind? He would lay it open on the examining table, so to speak, and see for himself. He would use his new instruments in the same manner that his predecessors had learned to use *their* instrument, the microscope. He would hover gently, behind the patient, and, by listening, look.

As he had with cocaine ten years ealier, Freud experimented on himself. If his office were a laboratory, he would do what he asked his patients to do, and he would perform on himself the same examination that he performed on them. So he lay on his couch and let his mind wander. Sometimes he monitored those wanderings to see where they led, and sometimes, just as important, he monitored those wanderings to see where they didn't lead—to see where he, or some part of him, refused to go.

And as was the case with cocaine, Freud's motive in testing this new instrument on himself wasn't entirely selfless. Freud was suffering. He was mourning for his father. Not that he knew he was mourning for his father, at least not consciously, at least not before lying down on the couch and subjecting himself to what he later called self-analysis. But as he lay there in his office, day after day, watching his thoughts, he found himself thinking about his father. He found himself thinking about the dream that had followed his father's death. He found himself thinking about dreams in general. He asked himself: If psychic energy proceeds along the pathways of the mind in the same manner as nervous energy proceeds along the pathways of the brain, shouldn't dreams be strictly deterministic?

And not just dreams. Shouldn't every thought—every psychic pulse, conscious or unconscious—carry some meaning? Just as Darwin's adaptations don't come from nowhere, just as they have to serve some purpose, so too our thoughts, whether conscious or unconscious, and so too the thoughts and words and actions to which they give rise. In Italy in 1898, for instance, Freud forgot the name of an artist and substituted the names of two other artists, a slip that he was able to trace to the remnants of the conversation he'd been having with a stranger and to his own wish to repress a sexual thought from entering the conversation. "How can I make that credible to anyone?" he asked Fliess in a letter. He knew how to try, anyway. Freud wrote it up in a paper.

He was moving far from the behavior of hysterics now; he was moving far from his own behavior, for that matter. He was beginning to address the behavior of everyone. And he felt that he was able to do so only because he had observed the workings of the human mind for himself, through the slow accumulation of empirical evidence in the sanctity of his own lab. This wasn't a laboratory like the one he'd entered in 1873 as a new medical student at the University of Vienna, and this wasn't a laboratory like the one he'd left in 1885 during his stay at the Salpêtrière. This was a laboratory like the one where Darwin had conducted his investigations, during his voyage on the *Beagle,* and then in the decades thereafter while he worked, day after day in his study, dissecting his specimens for meaning, building a theory about species. Could Freud do the same, only according to his own needs? Could he build a theory about one species?

"The intention of this project is to furnish us with a psychology which shall be a natural science," Freud had written in the opening sentence of his "Psychology for Neurologists." Other than mailing a copy to Fliess, Freud never released that project, but he did revive it. It appeared more or less intact—physiological components absent, psychological components present—in November 1899, in chapter 7 of his *Die Traumdeutung* (*The Interpretation of Dreams*), as part of an explanation of the method of treatment he'd been developing over the past few years in his private practice. In an 1894 paper he'd referred to this evolving method under various guises: "analysis," "hypnotic analysis," "psychical analysis," "psychological analysis," and "clinico-psychological analysis." Then on February 5, 1896—only two months after he himself had finally abandoned his "Psychology for Neurologists"—Freud sent off two papers, one in French, the other in German. In the French article, "Heredity and the Aetiology of the Neuroses," which appeared first, on March 30, he reported that he "owed his results to a new method." In the German, "Further Remarks on the Neuro-Psychoses of Defense," which appeared on May 15, he called it a "laborious but completely reliable method." In both cases, he chose the same term for this method, a term he from then on used virtually exclusively: "psycho-analysis." If Freud was right, psycho-analysis wasn't just a new form of therapeutic treatment, and not only a new instrument for investigation, but the tool that would allow him to develop a theory encompassing the results of those investigations: a new science.

III

THE TREMBLING OF THE DEWDROP

SIX

A DISCOURSE CONCERNING TWO NEW SCIENCES

"What, precisely, is 'thinking'?"

The question comes early in Albert Einstein's "Autobiographical Notes," an essay he wrote in 1946, at the age of sixty-seven. In anyone else's autobiographical notes, even another scientist's, the question might have been unusual, but for Einstein it was typical. More than typical, maybe: characteristic; a reflection of who he was at some fundamental level. Again and again over the decades after he had made a name for himself as the father of relativity in a new, non-Galilean sense of the word, in lecture after lecture and essay after essay, Einstein began not with an introduction to the subject at hand but with an overview of how he'd arrived at that subject, or of how scientists in general arrive at subjects in general. Even his most highly specialized scientific papers often opened with a brief account of how he'd come to reach the conclusions that followed. The reason for this narrative strategy wasn't simply that audiences—even audiences of his scientific peers—needed an expert guide in order to follow the thoughts of an Einstein (though no doubt sometimes they did). It was that for Einstein himself the results of science had become incomprehensible without an understanding of the processes that led to them. When he sat down to compose the essay that would be the closest he'd ever come to an autobiography—or, as he preferred to call it, "something like my own obituary"—he naturally chose to address not the concrete details of who, what, when, and where that usually make up a life story, but the abstract how: his growth as a thinker.

Similarly, Sigmund Freud, in a fragment of an essay he wrote in 1938 at the age of eighty-two, less than a year before his death: The quality of being conscious "remains the one light which illuminates our path and leads us through the darkness of mental life." But then, Freud had devoted all his life—at least in the nearly half a century since he'd abandoned the formal study of the brain—to the study of the mind. If for Einstein the thought process was a means toward a scientific end, for Freud it *was* the end. As he wrote that same year, "Every science is based on observations and experiences arrived at through the medium of our psychical apparatus." But, he continued, "*our* science has as its subject that apparatus itself"—that is, *anyone's* growth as a thinker.

It would be no exaggeration to say that all Einstein and Freud were doing was what philosophers had been doing forever, whether it was Plato positing the existence of ideal forms or Aristotle accepting as real only that which is unquestionably present: thinking about the way we think. Nor would it be an exaggeration to say that they were also doing what the practitioners of one modern variation on that same ancient discipline—natural philosophers, or, as they'd come to be known, scientists—

had been doing for the past few hundred years, upon closer and closer examinations of either the celestial or the cerebral: thinking about the way we think about nature. Still, Einstein and Freud's particular insistence on writing about the personal and idiosyncratic side of science—its subjectivity—presented a seeming paradox, coming as it did at the culmination of a centuries-long movement that supposedly valued most highly what its practitioners had trusted would be, in the end, objectivity.

That had been part of the appeal of the modern scientific method from the start. When a Galileo or a Leeuwenhoek concluded that, no, on closer examination, *this* is the way the universe works—the heavens or the human body—he was not only overturning age-old assumptions regarding what we know about nature. Knowingly or not, he was also changing *how* we know about nature. In the presence of incontrovertible evidence, the ultimate arbiter of what we believe about the universe was no longer an ancient authority or a distant deity. It was *you*—the individual investigator of nature. Or, more precisely, it was you as the surrogate for all the other individual investigators of nature who then could follow your lead, observing what you observed, measuring what you measured, and affirming or disputing your results.

And this method worked. Natural philosophers had looked at the heavens, and then they had looked farther, and through the telescope they saw new moons, planets, stars, and other systems of stars like our own Milky Way, or what came to be called galaxies. Natural philosophers had looked at the human body, and then they had looked deeper, and through the microscope they saw central nerve cells, fibers, and the neuronal points of contact between them all, or what came to be called synapses.

Worlds without and worlds within: Plato was right. For

thousands of years the universe had yielded one set of appearances. Now, through the addition to the human intellectual arsenal of a couple of new tools that for the first time ever extended one of the five senses, the universe had yielded an entirely different set of appearances. The telescope and microscope settled forever the question of whether the universe hid its secrets—of whether what was discernible to the unaided senses was merely a shadow of what was out there. Indeed, with every improvement, the telescope and the microscope revealed *more* of what was out there.

How much more? Einstein and Freud were hardly the only investigators of nature around the turn of the twentieth century to find themselves pushing up against the edge of scientific knowledge and asking, however idly, whether something had to give. And as it turned out, anyone wondering in the early weeks of 1896 about the future of scientific knowledge need have looked no further than the bizarre and shadowy image of Bertha Röntgen's hand on the front page of almost any newspaper. Not that this faint impression would have *seemed* like the start of a second scientific revolution—not at a historical moment that the two possible directions for the future of science came down to either the completion of knowledge or the addition of decimal places.

Still, this view of Bertha's more-or-less hand distinguished itself, even in an age of technological marvels. Only two months before Wilhelm Röntgen discovered X-rays, *The New York Times,* recounting the history of the still-new process of photography, called the previous twenty years those "which have produced more miracles of invention"—including the telegraph, the telephone, the phonograph, the motion picture and, above all, the electric light—"than did the whole of twenty preceding centuries." After Röntgen's announcement, the paper amended its view: X-rays were "unique."

Not quite and not for long. On March 1, 1896—two months to the day after Röntgen had sent his report and photographs to scientific colleagues and in direct response to one of those mailings—the French physicist Henri Becquerel found that if he wrapped a photographic plate in black paper, layered the paper with one of the phosphorescent substances he'd long been investigating, and stored the packet in a drawer for several days, the plate would show "silhouettes" that "appeared with great intensity." His conclusion: Even in total darkness, this phosphorescent substance—uranyl potassium sulfate, a compound of uranium—was emitting invisible rays. Later that same year, the Dutch physicist Pieter Zeeman repeated an experiment that the English physicist and chemist Michael Faraday had devised and attempted unsuccessfully more than forty years earlier to see if magnetism influenced light. It did, Zeeman could now confirm, thanks to the improvements in technology since Faraday's era, and almost immediately his fellow Dutch physicist Hendrik Antoon Lorentz arrived at a mathematical explanation that satisfied all the details of this result: Light is emitted by the movements of negatively charged particles within an atom—and this at a time when atoms themselves were still highly hypothetical and controversial, let alone particles *within* them. The year after that, 1897, the British physicist Joseph John Thomson stunned his audience at a "Friday evening discourse" at the Royal Institution by announcing that he'd proven the existence of these negatively charged particles—or electrons—by isolating them *outside* of their host atoms. And then the year after that, Marie Curie extended Becquerel's investigations on rays from uranium to other substances, in the process discovering radium and bestowing a new name on the overall effect: radioactivity.

And this was just physics. For some time, advocates for the establishment of other scientific disciplines had been wondering whether the lessons of physics in the material universe—the grand triumph of the Cartesian and Newtonian vision of matter and motion, motion and matter—would apply in less material realms. In particular, they'd been suggesting the effects of forces. Maybe not forces in the same quantifiable sense as physics—but then again, maybe so. At any rate, and in the same Newtonian tradition as gravity, *forces:* the somethings that put matter into motion, whether the matter is persons or planets, cells or civilizations.

The eighteenth-century economist Adam Smith had posited the influence on the marketplace of an "invisible hand." In the mid-nineteenth century, Karl Marx envisioned entire societies as organisms. Marx's collaborator Friedrich Engels himself once drew the comparison between the ideas of Marx and those of Darwin, whose theory of evolution through the process of natural selection quickly became the foremost exemplar of the force-of-nature model of science. If—a big if, but—*if* the physical sciences were drawing to a close at the turn of the twentieth century, maybe the nonphysical sciences would flourish next. "In 1894," the physicist Robert A. Millikan once recalled, "I lived in a fifth-floor flat on Sixty-fourth Street, a block west of Broadway, with four other Columbia graduate students, one a medic and the other three working in sociology and political science, and I was ragged continuously by all of them for sticking to a 'finished,' yes, a 'dead subject,' like physics, when the new, 'live' field of the social sciences was just being opened up."

Self-perpetuating rays routinely passing through supposedly solid objects that, at the level of the atom, are vastly more empty than not; psychological currents operating on the scale of civilizations and species, frustrating direct observation and measurement: These were the forces, or "forces," that increasingly were defying the old definitions of science. Were they, though, suggesting new frontiers in science? Were they suggesting new ways of thinking about the universe? On an individual basis, there would have been no immediate reason to believe so. Like Galileo's Jovian moons or Leeuwenhoek's jovial animalcules, each might have seemed an anomaly. But as was the case in the seventeenth century, if you put enough anomalies together, not only do they overturn assumptions regarding what we know about nature—what we see—but they begin to change *how* we know about nature: how we think about what we see.

The invisible, after all, was what the Scientific Revolution was supposed to have eliminated from rational discourse. Maybe not at first. Yes, Newton had succeeded in ridding the outer universe of invisible celestial spheres when he declared: "the Copernican system of the planets stands revealed as a vast machine working under mechanical laws here understood and explained for the first time." And yes, Descartes had succeeded in ridding the inner universe of invisible animal spirits when he declared: "I wish it to be considered that the motion which I have now explained follows as necessarily from the very arrangement of the parts, which may be observed in the heart by the eye alone, and from the heat which may be felt with the fingers, and from the nature of the blood as learned from experience, as does the motion of a clock from the power, the situation, and shape of its counterweights and wheels." Theirs, then, was a universe that still unmistakably moved, but now the causes of those motions were no longer remote and inaccessible.

Were they, though, present and accessible? Neither Newton nor Descartes wanted to venture a guess. *"Hypotheses non fingo,"* Newton famously declared, feigning no hypotheses as to how gravity works. Ditto Descartes: *"Cogito ergo sum,"* concluding that if he thinks, he therefore is. Both, however, acknowledged that the answer to the question of *how* the universe moves—what ultimately keeps the wheels of the celestial machinery turning or what ultimately distinguishes the cogs within the human brain from those of any other beast—surely had something to do with a Supreme Being. "He endures always and is present everywhere," Newton wrote of God in the General Scholium at the conclusion of the *Principia,* "and by existing always and everywhere he constitutes duration and space." And Descartes, having observed the dissections of various animals, including man, found that he could content himself that man did indeed possess something unique "as soon as I supposed God to have created a rational soul, and to have annexed it to this body."

A description of all things natural that in the end had to invoke the *super*natural, however, was hardly logically consistent. Newton himself had recognized the central limitation of his law of universal gravitation—that he had failed to explain how gravitation works. "That gravity should be innate, inherent, and essential to Matter, so that one Body can act upon another at a Distance thro' a Vacuum, without the Mediation of any thing else, by and through which their Action and Force may be conveyed from one to another," he wrote in a 1693 letter to the theologian Richard Bentley, "is to me so great an Absurdity that I believe no Man who has in philosophical Matters a competent Faculty of thinking can ever fall into it."

Yet fall into it any number of the philosophically minded did. Newton's contemporaries understood that his theory was a mathematical abstraction as to *what* the pieces of the universe do, not a guide to *how* the pieces of the universe do what they do, and his peers also recognized this argument regarding gravity as nothing more than the same one that natural philosophers had been making ever since Galileo himself made it, half a century earlier, in his 1638 *Discourses and Mathematical Demonstrations Concerning Two New Sciences.*

But over the decades, and then over the course of a century, and then over the course of a couple of centuries, the predictions possible through Newton's gravitation met not only with success but—what's the same but perhaps more important when it comes to converting a useful fiction into a conceptual fact, or "fact"—with no defeats, exceptions, or anomalies. Gravitation simply explained so much so well that a wish Newton had expressed in his author's preface to the first edition of the *Principia* wound up seeming prophetic. After summarizing how gravitation allowed him to account for "the motions of the planets, the comets, the moon and the sea," he added: "If only we could derive the other phenomena of nature from mechanical principles by the same kind of reasoning!"

We could, was the consensus. The generations of natural philosophers who succeeded Newton and Descartes credited them for framing the debates about gravity and consciousness, however fallaciously, and for forwarding the means with which anyone might begin to address these issues. But then they set out to answer what Newton and Descartes had not: Could the scientific method possibly encompass not only a universe that now unmistakably moved but one that *stayed in motion* entirely on its own, no matter how many levels of worlds upon worlds it ultimately contained? Could it describe a universe entirely free of what Newton himself called "occult"—from the Greek word for "hidden"—qualities?

In the heavens, the French mathematician Laplace set himself that challenge, and in his *Celestial Mechanics* he seemed to meet it. "You have written this huge book on the system of the world without once mentioning the author of the universe," Napoleon chided him. "Sire," Laplace replied, "I had no need of that hypothesis." Darwin did not explicitly set himself a similar challenge among the creatures of the Earth, but he seemed to meet it, anyway. "I would give absolutely nothing for the theory of Natural Selection," Darwin wrote to a friend in 1859, the same year he published *On the Origin of Species,* "if it requires miraculous additions at any one stage of descent."

Not just what it is; not just what it does: *how it does* what it does. "Every astronomer knows that there was only one secret of the universe to be discovered, and that when Newton told it to the world the supreme triumph of astronomy was achieved," the historian Charles Henry Pearson wrote in 1893—by which time, he went on, it was already apparent that Darwin might have "disclosed the other great mystery of the generation of life." Plato had posited the existence of hidden forms that cast the shadows we see, and then he had posited the existence of a vast mathematical framework that might support those forms; and now, more than two thousand years later, here it was: far more elaborate and far more intricate than any ordinary piece of machinery, perhaps, but ultimately, surely, as comprehensible as a clock.

And yet—and *yet:* For all those undeniable advances in

knowledge, the scientific method was still no closer to answering the two questions that had stopped Newton and Descartes in their ecclesiastical tracts: What is gravity? What is consciousness?

You couldn't really argue with the success of the cause-and-effect formulas of mechanics. What you could argue with instead was the reasoning that had led to it and therefore the overinterpretation that for two centuries had followed from it. "The Newtonian theory of gravitation, on its appearance, disturbed almost all investigators of nature because it was founded on an uncommon unintelligibility," Ernst Mach wrote in 1872. Now, he added, "it has become *common* unintelligibility." In his 1883 *The Science of Mechanics*—the book that Einstein cited as one of the greatest influences on his development of relativity and a work that Freud read admiringly at the behest of his and Mach's mutual friend Breuer—Mach updated Plato's parable of the cave for a cause-and-effect age: "A person who knew the world only through the theater, if brought behind the scenes and permitted to view the mechanism of the stage's action, might possibly believe that the real world also was in need of a machine-room, and that if this were once thoroughly explored"—that if we could exhaustively explain the world in mechanical terms—"we should know all." But, he argued, all we would actually know—all we do know—is the mechanical interpretation that we have imposed on nature, not nature itself. And so we should "beware lest the intellectual machinery, employed in the representation of the world on the stage of thought, be regarded as the basis of the real world." What lies beyond our sensory perceptions, beyond the strictest application of the scientific method, beyond physics, belongs—can only belong—to metaphysics.

Mach, in fact, was elaborating on an argument that the French philosopher Auguste Comte had begun constructing in the 1830s. Reflecting on the history of our relationship with nature, Comte had suggested that our intellectual development had progressed through three stages, each necessary. First had come everything leading up to the Scientific Revolution, the stage where natural effects are assumed to have not natural causes but *super*natural causes—spheres and spirits. This he called the theological stage. Next came the period of intellectual development begun by the likes of Newton and Descartes, who attempted to match natural effects with natural causes but in the end had overreached and in so doing had resorted to speculation. This Comte called the metaphysical stage.

So what's a scientist—and Mach always insisted he was approaching Newton's inconsistencies not as a philosopher but as an experimental physicist—to do? *Not* to try to grasp what isn't there. Rather, Comte suggested, natural philosophers should concentrate their investigations only on what *is* there, on the indisputably, provably present—on what he called the positive.

Positivism—as Comte conceived this third and final stage of our civilization's intellectual development and as Mach came to champion it—acknowledged the limitations of sense evidence, but it also recognized that sense evidence was all we had. Any information we receive about the universe can come only through one of our five senses. As inadequate or potentially misleading as sense evidence may be, it nevertheless should be not only a focus of our investigations, as it had been since the beginning of the modern scientific era, but the sole focus. In a lecture in 1872 that influenced the practitioners and philosophers of science for generations to come, the neurophysiologist Emil Du Bois-Reymond (he of the antivitalist "no other forces are in effect than the common physical and chemical ones" Berliner Physikalische Gesellschaft back in the 1840s) addressed the topic of "*Die Grenzen des Naturerkennens*" ("The Limits of Our Knowledge of Nature"). We can understand any natural phenomena, he argued, to which we can apply a mechanical model; in these cases, if our understanding is incomplete, we can conclude that the failure is ours and temporary. But for two riddles in particular no mechanical model was possible—not now, not ever: "What are matter and force, and how they are able to think." When confronting these questions, Du Bois-Reymond announced, we should respond not with the traditional, and provisional, *"Ignoramus"* ("We do not know") but with the more honest, if more humbling, *"Ignorabimus"* ("We shall never know").

What it is; what it does; but *not* how it does what it does: Aristotle was right after all. The world does not harbor a hidden reality—or if, as Plato held, it does, that deeper truth must remain forever unknowable and therefore as good as nonexistent.

This was the prevailing credo among scientists of central Europe by the turn of the twentieth century—the principle that guided natural philosophers as they either completed their inventory of nature or continued their assault on the vanishingly remote reaches to the right of the decimal point. This was the credo that informed the weekly meetings of the jokingly named Olympia Academy that Einstein and friends formed in Bern in the early years of the century, where they chewed over sausage and science and went for midnight hikes through nearby hills, all the while parsing and praising the writings of philosopher-scientists, including Mach and Poincaré. This was the credo of the zoology and physiology departments at the University of Vienna, where Freud first encountered natural philosophy, and of the Institute of Physiology in Vienna, where he spent his six most formative years of research. This was the credo that motivated the authors of a 1911 manifesto supporting the formation of the Gesellschaft für positivistische Philosophie (Society for a Positivistic Philosophy), whose thirty-three signatories included, in the most visibly prominent position—middle column, top—"Prof. Dr. Einstein, Prag" and "Prof. Dr. S. Freud, Wien." "What is the nature of force, of velocity, of the mind?"—these were the kinds of questions that the physicist Heinrich Hertz, consciously echoing Comte and Du Bois-Reymond and Mach, deemed "illegitimate," and these were the kinds of questions that Einstein and Freud found themselves, at first without their realizing it and often against their better judgment, attempting to address.

Einstein, for one, denied doing so, at least at first. "In your last letter, I find on re-reading, something which makes me angry: That speculation has proved itself to be superior to empiricism," Einstein wrote to his longtime friend and lifelong correspondent Michele Besso in August 1918. Einstein was objecting to what he felt was an assertion by Besso that he had resorted to nonpositivist premises in the development of relativity theory. "I find that this development teaches something else," Einstein went on, "that it is practically the opposite, namely that a theory which wishes to deserve trust be built upon generalizable facts." Among his "facts" Einstein included the equivalence between inertial and gravitational mass—something that, indeed, had been experimentally demonstrated by a Hungarian physicist, Baron Lóránt Eötvös, in 1890. At the time Einstein began the work that would culminate in the general theory of relativity, however, he knew nothing of that experiment. The only "fact" he knew—or, perhaps, "knew"—was what he'd glimpsed in his mind's eye while daydreaming in a patent office on an autumn day in 1907.

Freud didn't concern himself with philosophical niceties as much as Einstein did, and he privately strained at the restrictions of the positivist principles of the day. In December 1895, in the brief interim between his final effort to write his "Psychology for Neurologists" and his initial embrace of a purely "psycho-analytic" approach, he wrote to his friend Wilhelm Fliess that he hoped "you will not refrain from publishing even conjectures. One cannot do without people who have the courage to think new things before they are in a position to demonstrate them." It was, however, a kind of courage Freud thought he himself lacked. "I take no pride in having avoided speculation," he wrote ten years later, "but the material for my hypotheses has been collected by the most extensive and laborious series of observations." Only ten years after *that,* in 1915, in a paper on "The Unconscious," did he finally concede the speculative origins of his guiding hypothesis on that subject: "A gain in meaning is a perfectly justifiable ground for going beyond the limits of direct experience."

The limits of direct experience: This, if only in retrospect, was what an Einstein or a Freud at a crucial juncture in his investigations had to respond to: a lack of evidence at the heart of his field. Einstein had focused first on the problem of absolute space, then refocused on absolute time—and realized that the solution to his problem lay in the *absence* of absolute time. Freud had focused first on the problem of mind and body, then refocused on consciousness alone—and realized that the solution to his problem lay in an *absence* of consciousness. No wonder they'd found themselves, however unwittingly, violating the principles of positivism—a view of science, after all, that granted validity only to evidence that was irrefutably *present*. When they'd run out of evidence, they'd looked harder, and what they wound up discovering wasn't new evidence but a new way of looking at old evidence, a new way of seeing that encompassed a subtle intellectual distinction that everyone else had missed: a shift not in perception—not in the apprehension of what was available to the senses—but in conception.

A visitor to Röntgen's laboratory once asked what he had thought when he first observed the spooky glow and discovered its see-through properties. He answered, proudly, "I did not think; I investigated." So it was with Einstein and Freud, at first. They investigated, and to the extent that they did think, it was in keeping with the scientific tradition of trying to let the evidence lead them where it may, not the other way around. The evidence they were investigating, moreover, was the same evidence that their contemporaries were investigating. Then the lack of evidence they found themselves investigating was the same lack of evidence that their contemporaries were investigating. Yet Einstein and Freud wound up venturing where their contemporaries did not because at a certain point, they *didn't* investigate. They thought.

They reconceived the problem. Confronting a lack of evidence, they did what their training and their beliefs directed them and what any number of scientific anecdotes similar to the one about Röntgen's dogged investigations into a mysterious glow instructed them—what just about every bias in their being told them explicitly—*not* to do. They feigned a hypothesis.

Once, Einstein met Mach. It was in the autumn of 1913, and Einstein was in Vienna to deliver an address before the Congress of German Scientists and Physicians. He took the opportunity to make a pilgrimage to the suburbs of Vienna and pay his respects to the philosopher who had exerted such a great influence on his own philosophical development. Mach, for his part, was eager to meet the man behind a new idea of relativity that, according to reports that had reached Mach, owed a debt to his own writings. When Einstein arrived at Mach's apartment, he found a physically diminished figure, gray and wild of beard and suffering from the paralysis that had crippled him some dozen years earlier. Mach was still intellectually sharp, though, so Einstein seized the chance to ask the question that had been bothering him for some time now: If a physicist made a speculative leap that allowed him to establish a relationship among apparently disparate phenomena in a way that observation could then corroborate, would Mach agree that such an approach was still positivistic? Mach replied that he supposed, with some effort, he could imagine such circumstances. Einstein went away satisfied; Mach, apparently, remained less so, for in a preface to a posthumously published book of his, he disavowed any influence on Einstein's work and in fact promised a philosophical refutation of relativity's premises (which he didn't live long enough to write).

Perhaps not coincidentally, it was only a few months after the publication of this book of Mach's that Einstein finally allowed himself to begin to acknowledge what was becoming increasingly apparent, that, in retrospect, he'd been straining to deny. One year, 1921, Einstein was saying about relativity at a lecture in London, "This theory is not speculative in origin." The next year, at a lecture in Paris, he was reversing himself: "Mach's system studies the existing relations between data of experience: for Mach, science is the totality of these relations. That point of view is wrong, and in fact what Mach has done is to make a catalogue, not a system. To the extent that Mach was a good mechanician," Einstein pressed on, "he was a deplorable philosopher."

To illustrate the logical limitations of positivism, Einstein could cite his own general theory of relativity. Here was an explanation of gravity that corresponded with observations; but then, so did Newton's, if observations of what was irrefutably present were *all* you had to build a theory on. "The fictitious character of the fundamental principles is perfectly evident from the fact that we can point to two essentially different principles"—Newton's action-at-a-distance notion of gravity and his own general relativity—"both of which correspond with experience to a large extent," Einstein said in his Herbert Spencer lecture at Oxford in 1933. "This proves that every attempt at a logical deduction of the basic concepts and postulates of mechanics from elementary experiences is doomed to failure."

With hindsight, positivism and mechanics had been a perfect match. If you (even unconsciously) assumed that you were describing a universe that was nothing but motion and matter, restricting your investigations to increasingly close studies of matter in motion wouldn't merely suffice—it would tend to reinforce the mechanistic view. Mach himself had warned of the dangers of regarding the machine-room of the mind as the basis of reality, rather than as the basis of the mechanical view of the world that our powers of observation impose on it, but the methodology he advocated for investigating the world wound up ensuring only the discovery of more and more mechanisms.

Not that Mach's skepticism about Newtonian mechanics wasn't useful. In arriving at his special theory of relativity, Einstein had found inspiration in Mach's questioning of mechanics as the basis of all scientific thought; Mach's example helped prompt him to consider alternative interpretations of the available evidence, an intellectual debt that Einstein acknowledged to the end of his life. But Einstein wouldn't have arrived at those alternative interpretations—those two essentially unverifiable postulates: the extension of the relativity principle beyond mechanics and into electrodynamics; the constancy of the speed of light—without venturing where Mach forbade. On that fateful May evening in Bern in 1905, as Einstein eventually wrote in his "Autobiographical Notes," the essay he considered his obituary, what he found himself confronting wasn't fresh evidence but an *absence of evidence:* "By and by I despaired of discovering the true laws by means of constructive efforts based on known facts." As Einstein wrote to Besso regarding Mach's brand of positivism, it can "only exterminate harmful vermin"; it "cannot give birth to anything living."

If the methods of positivism ensured a mechanistic interpretation of the world—access to what was demonstrably present—speculation allowed something else—access to what wasn't demonstrably present: the invisible. The whole of the electromagnetic theory that would lead Einstein to his special theory of relativity, in fact, had begun with a speculative leap. Back in 1821, when the British physicist Faraday had created the first dynamo by placing a magnet on a table and sending an electrical impluse through a wire that was hanging over it, the wire had begun twirling as if the electricity and the magnetism were together creating a whirlpool.

"Do you see?" Faraday called to his brother-in-law. "Do you see? Do you see?"

His brother-in-law saw, all right, but more to the point was that Faraday had seen it first—before he'd even switched on the battery. Formally uneducated, Faraday possessed an intuitive gift of visualization, and what he imagined in his mind's eye when he looked at a battery generating electricity or at a magnet emanating magnetism wasn't what every other physicist of the day imagined—straight lines of force. Instead, he "saw" circles that would interact in a way that would account for a whirlpool effect. "Fields," he called these circles, and on April 10, 1846, in an impromptu lecture at the Royal Institution in London, Faraday suggested that the circular fields of force he'd been using for the past quarter of a century to help him visualize the otherwise invisible interactions between electricity and magnetism had begun to seem real. After all, you *could* almost see them. You could cover a magnet with a sheet of paper, sprinkle the sheet with iron filings, and observe with your own eyes the filings immediately arranging themselves according to patterns emanating from the poles of the magnet. Faraday's audience that day, sated by a century and a half of visible cause-and-effect Newtonian mechanics, was skeptical. But once James Clerk Maxwell's four equations of electromagnetism provided the hypothetical fields with a secure mathematical foundation, other physicists had no choice but to follow Faraday in his acceptance of this alternate vision of the world. By the time Einstein took up the problems of electromagnetism, he not only accepted fields, he embraced them. "This change in the conception of reality," he wrote on the one hundredth anniversary of Maxwell's birth, "is the most profound and fruitful one that has come to physics since Newton."

Einstein didn't claim to be original in having made a speculative leap in working toward a theory in physics. In fact, the more he thought about it, the more he recognized it as a leap that had been present at the birth of the scientific method.

Hadn't Johannes Kepler arrived, after seven years' labor, at a geometric form to describe the orbit of Mars—an ellipse—not by generalizing from the equations he'd derived from Tycho Brahe's observations but through conjecture? "The orbits were empirically known," Einstein explained, "but their laws had to be guessed from the empirical data. First he had to make a guess at the mathematical nature of the curve described by the orbit, and then try it out on a vast assemblage of figures. If it did not fit, another hypothesis had to be devised and again tested. After tremendous search, the conjecture that the orbit was an ellipse with the sun at one of its foci was found to fit the facts."

And Kepler was only the beginning, the model of the scientific method as succeeding generations had come to (mis)understand it. What the likes of Ernst Mach never realized, Einstein once wrote to his friend Besso, was "that this speculative character belongs also to Newton's mechanics, and to every theory which thought is capable of." Einstein's point wasn't that a Röntgen, for instance, *should* have been thinking in addition to investigating. It was that he *had* been thinking, just as Faraday had been thinking or Maxwell had been thinking. Einstein acknowledged that making a distinction between conceptions and perceptions was the "metaphysical 'original sin,'" and he also acknowledged that he himself had committed it. But he wasn't alone. It was, he would argue throughout his life, a sin that every scientist committed. "I believe," he wrote in 1950, at the age of seventy-one, only five years before his death, "that every true theorist is a kind of tamed metaphysicist, no matter how pure a 'positivist' he may fancy himself." Or, as Einstein cautioned in the opening lines of his Oxford lecture, "If you want to find out anything from the theoretical physicists about the methods they use, I advise you to stick closely to one principle: don't listen to their words, fix your attention on their deeds."

If Einstein readily acknowledged he wasn't original in having made a speculative leap, he also realized he *could* be inspirational. It was, after all, *his* synthesis of Faraday and Maxwell's electromagnetic theories with Newton's mechanics—not *their* reconception of electricity and magnetism in terms of fields—that had come to symbolize the culmination of the classical era in physics. Einstein could do for his peers what Mach had done for him—guide them in exterminating harmful philosophical vermin. And he could also do for his peers what Mach had *not* done for him—guide them in giving birth to something new. Einstein's self-assigned mission, then, wasn't to change the way that natural philosophers did what they did. It was to change they way they *thought* they did what they did. In effect, Einstein's example gave them permission to do what they'd been doing all along.

His fellow physicist Werner Heisenberg recalled a long conversation with Einstein on this topic one night in 1926, at a time when Heisenberg mistakenly believed that Einstein was, like himself, still a pure positivist. Heisenberg was discussing the problem currently occupying him, the difficulty of using empirical evidence to describe with any certainty the orbits of electrons within an atom—a phenomenon, after all, that was unobservable.

"But you don't seriously believe," Einstein protested, "that none but observable magnitudes must go into a physical theory?"

"Isn't that precisely what you have done with relativity?" Heisenberg answered in surprise. "After all, you did stress the fact that it is impermissible to speak of absolute time, simply because absolute time cannot be observed; that only clock readings, be it in the moving reference system or the system at rest, are relevant to the determination of time."

"Possibly I did use this kind of reasoning," Einstein said, "but it is nonsense all the same." Observations may well be useful and informative and instructive in constructing a theory, Einstein added, and in his initial desire to defend relativity against his own suspicions that he wasn't adhering to the tenets of positivism, he probably overemphasized their importance. "But on principle," he continued, "it is quite wrong to try founding a theory on observable magnitudes alone. In reality the very opposite happens. It is the theory which decides what we can observe."

Or: It is a new conception that allows new perceptions. It is relativity that allows the apprehension of slowed time or curved space. It is the dynamic unconscious that allows the apprehension of the repressed.

But: *None of it amounts to anything without evidence.* The basic premise of the scientific revolution hadn't changed: You alone weren't enough, even if you were an Albert Einstein or a Sigmund Freud.

In this way, Einstein and Freud never broke faith with positivism. They didn't simply leap to conclusions; they didn't only hypothesize. They hypothesized, and then, as need be, depending on the evidence, they revised, until the hypothesis matched observations. Freud, in a 1915 letter, characterized scientific creativity as a "succession of daringly playful fantasy"—the speculative leap that he was then finally ready to concede he'd made in his initial identification of the unconscious—"and relentlessly realistic criticism." He later elaborated on the ultimate goal of speculation in scientific thought: "Its aim is to arrive at correspondence with reality, that is to say with what exists outside us and independently of us." Likewise Einstein, when he outlined in his Oxford lecture a speculative approach to science that "was by no means the prevailing one in the eighteenth and nineteenth centuries," made sure to add that "all knowledge of reality starts from experience and ends in it." Speculation helps bridge the abyss between "the fundamental concepts and laws on one side and, on the other, the conclusions." But: "Experience remains, of course, the sole criterion of the physical utility of a mathematical construction."

Einstein, for instance, had offered three predictions as a way to match his mathematical constructions with experience. He might have offered the expansion of the universe as a fourth, but even though he didn't, its discovery was generally accepted within the community of physicists as just that: further validation. As for Freud, if asked where his evidence was, he might have answered: Where wasn't it? It was in slips of the tongue, in jokes, in dreams, in wishes and sighs, in phallic symbols and waking fantasies. Once you learned to identify the dynamic unconscious in all its guises, it was seemingly in action everywhere, in everyone, all the time.

"The new Columbus," Einstein was often called. At the Royal Society meeting where Arthur Eddington announced the results from the 1919 solar eclipse expeditions, Joseph John Thomson characterized general relativity as "not the discovery of an outlying island, but of a whole continent of new scientific ideas of the greatest importance to some of the most fundamental questions connected with physics." The following month, a Johns Hopkins University professor, J. S. Ames, used his presidential address before the United States Physical Society to alert the organization's membership to the implications of general relativity: "I have all the enthusiasm of the discoverer of a new land, and feel compelled to describe to you what I have learned."

The same sort of metaphor often applied to the unconscious. When it wasn't being compared to an iceberg or an island, or at least those parts that remain mostly below the water's surface, the unconscious was "the inner Africa." Freud himself once seized on this analogy. Although he was describing, in a letter to Fliess, his own personality, his choice of metaphor was revealing: "I am nothing but by temperament a conquistador."

But the analogy, however seemingly appropriate for an era when the geographic Age of Discovery that had dawned five centuries earlier was drawing to its Earth's-landmass-determined close and however resonant in an era when the scientific age of discovery that had dawned three centuries earlier was also quite possibly drawing to its nature's-secrets-determined close, went only so far. Even though Einstein's or Freud's work brought new phenomena to view—and even if many of those phenomena were of a type that nobody previously could have imagined—they were not, after all, phenomena that hadn't been before our eyes all along. It wasn't as if the problem in perceiving them was that they had been as inaccessible as a couple of continents on the other side of the globe. And it wasn't as if the problem in perceiving them was that they had been too distant across the heavens or too minute within the human body to detect without the help of optical instruments. They had been *right there:* on a train platform across town, or in the words of a parent. What was challenged by the discovery of slowed time or curved space, of resistance or the repressed, were traditional assumptions regarding not perspective—not how far or how deep we can see—but perception: how we see, period.

X-rays weren't longitudinal vibrations in the ether, of course, but what they were nonetheless wound up anticipating and then reinforcing Einstein's ideas about relativity: They were a form of light. An "invisible light," as popular accounts called it—a term that seems paradoxical unless you happened to know how many phenomena the word "light" could encompass. As one newspaper report on X-rays said, "men of science have been long aware that there are rays of light imperceptible to the senses"—beyond red or violet in the spectrum of light. Early in the nineteenth century, William Herschel (the same astronomer who marveled that the deeper he looked into space, the farther back he was seeing in time) found that if he used a prism to break a beam of white light into its constituent colors, thermometers he placed above and below the extremes of the spectrum would register higher and lower temperatures where no light appeared to shine. Later experiments showed that colors of light correspond to the lengths of their waves—longer to shorter going from red to violet. And only seven years before Röntgen's discovery of X-rays, Hertz used Maxwell's field equations to prove the existence of very long waves of electromagnetism, which came to be called radio waves. As was becoming increasingly apparent—in the words of one post-X-ray-discovery headline—"Darkness Need Not Necessarily Imply an Absence of All Forms of Light."

"The great majority of people never bother themselves about what light is," read one account; "they take it for granted that it is one of those things that no one can find out, and therefore never give it a thought." Now, they did. A new conception of light and of seeing and of our relationship with light and seeing was emerging. The same report on X-rays that cautioned potential sinners against thinking they would be safe if they pulled down the blinds did so explicitly because our "radiations are traveling about somewhere in the world." Your image, in other words, isn't you; you aren't your image. The image of you that others see is *separate* from you. Technically, when people "see" you, what they're seeing is light signals carrying information about you. This way of thinking about seeing might have been implicit for natural philosophers who had been debating since the publication of Newton's *Optics* whether light travels in waves or particles and the invention of photography in the first half of the nineteenth century had gone a long way toward disseminating and reinforcing this way of thinking about seeing. But the discovery of X-rays helped make it explicit: Just as images present themselves through the propagation of waves of light from an object to the photographic plate or from a Hittorf-Crookes tube *through* an object to a photographic plate, so they also present themselves to the eyes of an observer.

To the eyes of, say, someone seeing a train pull into a station at seven o'clock.

"We went from the lecture hall as in a dream," wrote a correspondent of the *Manchester Guardian* after a lecture on relativity by Einstein in Vienna. But if Einstein were right, that's how they'd come into the hall, too. "The reader," Eddington wrote in *The Contemporary Review* only a month after the release of the results of the eclipse expeditions, "must remember that the space that is warped is actually the space of his perception." Through force of habit, if not necessity for survival, we have come to think that our minds are in direct contact with the world outside us, that what we see is what's irrefutably there according to some absolute standard. Not so. "So far from this being true," Eddington continued, "two of the most essential features in our mental picture of the external world—viz., space and time, are not actually in the external world." So where are they? They "are introduced into the picture in the course of transmission through sense-channels to our brains."

There was nothing to be done about this limitation, of course. It simply came with being a member of the species. "It will be seen that we are not fully equipped by our senses for forming an impersonal picture of the world," Eddington wrote in a 1920 book on relativity. "Perhaps if we had been endowed with two eyes moving with different velocities our brains would have developed the necessary faculty; we should have perceived a kind of relief in a fourth dimension so as to combine into one picture the aspect of things seen with different motions." *The Independent* echoed that regret: "It does not matter that we cannot 'see' a figure in four dimensions even with our mind's eye. Actually we cannot see any figure of more or less than two dimensions: we have to take the others on faith." Or, as Eddington wrote, "There is no illusion in these phenomena, except in so far as all our acquaintance with nature is illusion."

The same implications emerged through Freud's work, only Freud was concerned with the creation of the illusion itself: the stories we tell ourselves and one another about ourselves without even realizing that we're doing so. Through the technique of psychoanalysis, Freud argued, we for the first time had access to the ways the unconscious disguises our intents, to what those intents are, and to how those intents nonetheless manifest themselves in our daily, and nightly, lives. "Psychoanalysis," Freud wrote in a letter after the outbreak of war in

1914, "has inferred from the dreams and parapraxes of healthy people, as well as from the symptoms of neurotics, that the primitive, savage and evil impulses of mankind have not vanished in any of its individual members, but persist, although in a repressed state, in the unconscious (to use our technical terms), and lie in wait for opportunities of becoming active once more."

In this respect, too, X-rays proved instructive. They didn't allow us to see thoughts, of course, but one way they worked nonetheless wound up anticipating and then reinforcing Freud's ideas about how the unconscious worked. They illustrated a delay between—the apparent lack of immediate contact between—a cause and an effect. Even in the first months after Röntgen made his discovery public, reports were beginning to circulate that X-rays somehow seemed to be associated with sunburned flesh; skin stripping off and hair dropping out; the inflammation of eyelids, upper lips, and facial skin, especially in the experimenters who had exposed themselves to the rays repeatedly and over long periods; perhaps, in the case of one doctor, death. Yet even as late as 1903—seven years after Röntgen's announcement and long after such horrifying reports had multiplied in number and morbidness—one doctor reported his frustration in trying to cure a patient of cancer through exposure *to* X-rays: "Why, I have a patient whom I've been trying to burn for a year, and he simply won't burn."

Suffice to say that it wasn't long before the notion of something extraordinarily powerful and possibly destructive being absent yet active didn't seem illogical. What Freud had to offer, as he acknowledged on several occasions, had long been evident in the works of artists. True, writers may "show only slight interest in the origin and development of the mental states which they portray in their complete form," Freud once wrote. "But creative writers are valuable allies and their evidence is to be prized highly, for they are apt to know a whole host of things between heaven and earth of which our philosophy has not yet let us dream. In their knowledge of the mind they are far in advance of us everyday people, for they draw upon sources which we have not yet opened up for science." This, then, was Freud's self-appointed task: to open up the unconscious for science: to take what had long been at best an unthinking subtext of a larger discussion among philosphers, psychologists, and artists and to elevate it to the level of text.

By now, the case histories that Freud once thought might lack "the serious stamp of science" had acquired it, precisely for this reason. They were not, as Freud had feared when he began submitting them as evidence in his and Breuer's *Studies on Hysteria,* anecdotal—that is, one circumstance that might or might not support the larger point. They *were* the point. The dreams in the *Interpretaion of Dreams* in 1900, the slips of the tongue and lapses of memory in *The Psychopathology of Everyday Life* in 1901, and the aggressive stabs at humor in *Jokes and Their Relationship to the Unconscious* in 1905: These were crucial, the heart of the argument, the evidence from his laboratory. They were Darwin's finches.

Freud's work didn't burst upon the public in one brilliant burst, as Darwin's had in 1859, and as Einstein's did in 1919. Freud privately complained at the lack of the immediate reponse to the publication of *Interpretation of Dreams* (in late 1899; the publication date anticipated the new year), but his wait for recognition in fact wasn't long. The Wednesday Psychological Society meetings he held in his apartment and office at Berggasse 19 (where he'd moved his family in 1891) helped spread the word among his peers about a new instrument for examining the human mind and what it was revealing about human nature. In 1902 Freud received the honorary title of Ordinary Professor from the University of Vienna, an occasion he described as "a shower" of "congratulations and flowers." In 1908 the first congress of "Freudian Psychology" was held in Salzburg. The following year, Freud took his first trip to America to lecture at Clark University in Worcester, Massachusetts, where he found his teachings even more enthusiastically embraced than at home.

Often Freud's assertions were sensational or at least disturbing. In 1905, in *Drei Abhandlungen zur Sexualtheorie* (*Three Essays on the Theory of Sexuality*), he finally made explicit to himself, and for the first time to the public, a suspicion he'd been harboring for the better part of a decade: that children are sexual creatures. It was a claim that for much of his European and American readership would have violated the current conception of childhood as a pure state of being. But that vision of innocence itself had arisen in part as a reaction against an earlier conception of children as merely miniature adults—and therefore, by implication, sexual in some way. Perhaps the time had come not to endow the subject of childhood with whatever predispositions the current cultural climate favored but, instead, to evaluate the evidence on its own merits and draw conclusions about this species once and forever.

And that's what Freud offered. That's where the strength of his arguments lay, not just on the subject of sexuality but all the others he was addressing in the early years of the century: evidence. He had compiled the evidence of individual cases, but through the steady accumulation of evidence over the years Freud offered a compelling argument that these individual cases not only were representative of other cases he'd encountered but that they were symptomatic—that they offered insights into motives and behaviors and drives that were (inner) universal.

Just as the invention of photography and the discovery of X-rays had allowed what was implicit about the nature of seeing to become explicit—that light carries information to our eyes—and just as the example of Einstein was allowing what was implicit about the nature of thinking about science to become explicit—that a speculative leap is inevitable—so the exploration of the unconscious took what was implicit about the nature of thinking itself and regard it as the object of our investigations: that we are at the mercy of our own impulses. What Eddington wrote about the implications of relativity was equally true of the implications of psychoanalysis: "If our common view of nature is a dream, still our business is with the fabric of our vision. The discovery is epoch-making, and much that appeared inexplicable in the dream-world is now traceable to its source. *But the dreamer goes on dreaming*."

And that was that. The key insights of Einstein and Freud were indeed epoch-making, and so their epochs were made. These two turn-of-the-twentieth-century natural philosophers had each confronted a problem that was defeating their contemporaries, had figured out that the solution lay not in a new perception but in a reconception of the problem, had used that new conception to find evidence that turned out to be not so much new perceptions as old perceptions in a new light; and now the appearance of those old perceptions in a new light had led the two of them, their scientific peers, and even the public to a new conception of perception altogether. Einstein and Freud had made their epochs, and now their epochs continued, as epochs must, without their makers.

By the time Freud and Einstein met early in 1927, they were legends. Befitting their differences in age and health, it was (in Freud's words) the "cheerful, full of himself and agreeable" forty-seven-year-old Einstein who called on the cancer-afflicted, prosthetic-palate-impaired, all-but-deaf-in-his-right-ear seventy-year-old Freud. Although their two hours together passed pleasantly enough, they found little common ground, with the possible exception of the one they continued to explore, in occasional correspondence, for the following decade: the way their legacies were already getting away from them.

"Everything is relative" became a common interpretation of Einstein's labors. A 1920s motion picture short, for instance, supposedly explained relativity by showing a clock seeming to move slowly from the point of view of a schoolboy taking a test. But such commonplace "explanations" only confused an emotional state with a condition of physics—a confusion that Einstein himself unfortunately helped promulgate. At the end of his first public writing after the eclipse announcement—an article originally published in the November 28, 1919, edition of the *The Times* of London, then widely reprinted worldwide to satisfy the tremendous public curiosity about this ostensible slayer of Newton—Einstein added this joke: "Here is yet another application of the principle of relativity for the delectation of the reader: today I am described in Germany as a 'German savant,' and in England as a 'Swiss Jew.' Should it ever be my fate to be represented as a *bête noire,* I should, on the contrary, become a 'Swiss Jew' for the Germans and a 'German savant' for the English."

The real lesson of relativity, however, was exactly the opposite of "Everything is relative": Some things aren't. What's relative are observations between two physical systems. What's not is the single set of laws that describes the relationship between any two physical systems in the universe. The German physicist and early Einstein advocate Max Planck first applied the term "relativity theory" to Einstein's work in 1907; Einstein himself publicly referred to his work only as the "so-called relativity theory" until 1911, when he finally dropped the qualifier and accepted the common usage. His personal correspondence during this period indicates that he preferred the term *Invariantentheorie,* or invariant theory, which would have been a far more accurate reflection of his intentions: one set of explanations for any relationship in physics.

Einstein thought that to generalize from a physical theory to a statement of the human condition—to draw from a mathematical proposition an existential conclusion—"is not only a mistake but has something reprehensible to it." Freud agreed, no doubt in part because his own work also seemed to lend itself to observation-is-in-the-eye-of-the-beholder misinterpretations. As early as 1921, just two years after the results of the eclipse expeditions made Einstein's name a household word, Freud not only came to Einstein's defense but exhibited a sensitivity to the misnomer that had attached itself to Einstein's work: "The knowledge that has been so very recently acquired of what is called the theory of relativity has had the effect upon many of those who admire without comprehending it of diminishing their belief in the objective trustworthiness of science." In 1933 Freud amplified this argument: "No doubt there have been intellectual nihilists of this kind before, but at the present day the theory of relativity of modern physics seems to have gone to their heads," he wrote. "According to this anarchistic doctrine, there is no such thing as truth, no assured knowledge of the external world. What we give out as scientific truth is only the product of our own needs and desires, as they are formulated under varying external conditions; that is to say, it is illusion once more. Ultimately we find only what we need to find, and see only what we desire to see. We can do nothing else. And since the criterion of truth, correspondence with an external world, disappears, it is absolutely immaterial what views we accept. All of them are equally true and false. And no one has a right to accuse anyone else of error."

As was the case for Einstein, what Freud sought in his work was exactly the opposite of the meaninglessness he bemoaned here: inflexible principles that could account for the unconscious in all its flexibility. Whether he succeeded was a matter of debate, and regarding psychoanalytic theory Einstein himself wavered, as he wrote Freud on the latter's seventy-fifth birthday, between "belief and disbelief." But, Einstein added, he was reading Freud's work with a friend every Tuesday evening and greatly admiring its "beauty and clarity. Except for Schopenhauer there is, for me, no one who can or could write like that." The following year Einstein thanked Freud "for many a beautiful hour which I owe to the reading of your works," then added, "I always find it amusing to observe that even people who regard themselves as 'unbelievers' with regard to your teachings, offer such little resistance to your ideas that they actually think and speak in your concepts the moment they let themselves go."

Such was the case with Einstein. Even if he possibly counted himself among the "unbelievers," Einstein's appreciation of Freud's work was genuine, and when the League of Nations' Institute for Intellectual Cooperation enlisted him to engage in a public correspondence with another world-famous figure, Einstein immediately approached Freud. In answer to the question *Why War?,* the title of their published exchange, Einstein wrote: "Because man has within him a lust for hatred and destruction. In normal times this passion exists in a latent state, and it emerges only in unusual circumstances; but it is a comparatively easy task to call it into play and raise it to the power of a collective psychosis." To which Freud, who privately thought the project "tedious," could only respond, "you yourself have said almost all there is to say on the subject."

Freud, in fact, thought that he entered into any exchange with Einstein at a disadvantage. "The lucky fellow," Freud wrote to a friend regarding Einstein shortly after their evening together, "has had a much easier time than I have. He has had the support of a long series of predecessors from Newton onward, while I have had to hack every step of my way through a tangled jungle alone. No wonder that my path is not a very broad one, and that I have not got very far on it." Two years later, Freud echoed this sentiment in a fiftieth-birthday greeting to Einstein, calling him a "happy one." Einstein wrote back: "Why do you emphasize happiness in my case? You, who got into the skin of so many people, and indeed of humanity, have had no opportunity to slip into mine." Freud answered that he'd simply assumed Einstein was happy because he "could work at mathematical physics and not at psychology where everyone thinks they can have a say."

Some envy on Freud's part was understandable. When Einstein attended the 1931 Hollywood premiere of *City Lights,* the movie's star and director Charlie Chaplin offered him this explanation of the crowd's response: "They cheer me because they all understand me, and they cheer you because no one understands you." Freud could have said the same of himself in relation to Einstein, except sometimes for the cheering part.

Freud famously—or infamously—traced the opposition of his "unbelievers" to a resistance against a "third blow to man's self-love." The first blow, he wrote in 1917, had been the Copernican removal of Earth from the center of the universe to become merely one more planet. The second was the Darwinian removal of man from a privileged position among the creatures of the earth to become merely one more species. The third blow was his own: "that *the ego is not master in its own house.*" "No wonder, then," he went on, "that the ego does not look favourably upon psycho-analysis and obstinately refuses to believe in it." Einstein, meantime, didn't need to make such claims for himself; others were making them for him. A month after the eclipse announcement, Eddington declared Einstein's work to be "comparable with, or perhaps exceeding the advances associated with Copernicus, Newton, and Darwin."

Whatever the merits of Freud's logic—essentially, that if you don't like psychoanalysis it's because you can't handle the truth—it wasn't an argument that was going to win him many converts. More to the point, it didn't address the possible merits of his opponents' argument: that compared with a science such as physics, psychoanalysis didn't measure up—literally; that it wasn't a science because its results weren't quantifiable and subject to outside comfirmation.

Einstein's campaign to change the way natural philosophers thought they did what they did was effective. Logical positivism this new methodology—or, perhaps, this new recognition of the existing methodology—was called by the Vienna Circle, the philosophical movement that began as a discussion among mathematicians, physicists and philosophers before the Great War, led to a formal declaration of principles in 1929, and culminated in 1934, the publication of Karl Popper's *The*

Logic of Scientific Discovery. For Popper in particular and the logical positivists in general, the formulation of Einstein's theories represented how science *does* work and *should* work. Whether theories such as Einstein's ultimately reveal an immutable truth about the universe, as Plato held or simply make more explicit the relationship between objects in the tradition of Aristotle continued to be a matter of debate. But an interplay of Platonic and Aristotelian methods—of speculation corroborated by close observation—became the scientific standard.

It was this model that Freud always insisted he followed. Yes, Freud acknowledged, the unconscious "is a hypothesis, and science"—he immediately made sure to add—"makes use of many." "And here is the opportunity to learn what we could not have guessed from speculation, or from another source of empirical information," he wrote in 1912, in "A Note on the Unconscious in Psychoanalysis," "that the laws of unconscious activity differ widely from those of the conscious." It is therefore the hypothesis of the unconscious that enables psychoanalysis to take its place as a "natural science like any other."

But was it like any other? What was this unconscious that Freud claimed he discovered? Where was it? To say, based on observations of the interplay of causes and effects, that the unconscious simply *had* to be there was to risk repeating the fallacies that attended the "creation" of gravitation and the ether—or of celestial spheres and animal spirits, for that (non)matter. At one typical symposium, "Is the Conception of the Unconscious of Value in Psychology?," at the Joint Session of the Mind Association and the Aristotelian Society at Manchester, England, in 1922, one critic voiced this common criticism: "to ask us to think of something which has all the characteristics of a wish or a feeling except that it is not conscious seems to me like asking us to think of something which has all the attributes of red or green except that it is not a colour." Or, in the words of another critic, "The scientific level of Freud's concept of the unconscious is exactly on a par with the miracles of Jesus."

It got worse. In the same influential book in which Popper held up Einstein's theories as the model of how science does and should work, he also held up Freud's theories as the model of how science *doesn't* work and *shouldn't* work. For Popper, the crucial test of a theory wasn't whether you could conduct an experiment to prove it valid. Such a definitive judgment, he maintained, was unrealistic. Rather, he argued, the crucial test of whether a theory was scientific was what he called its falsifiability: Could you conduct an experiment that would prove a theory *wrong?* Einstein, in fact, always maintained that if relativity failed any one of the tests he had devised for it, the entire theory would collapse. A physical theory was, he agreed, an all-or-nothing proposition.

It's undeniable that Freud wanted psychoanalysis to be a science and that for him the unconscious was not a mere "figure of speech." He insisted on both repeatedly. Even in the final year of his life, having fled the Nazis for the safety of suburban London, slowly dying of cancer, a prosthetic device holding his face together, barely able to speak, he nonetheless made sure to identify himself in an interview with the BBC as the founder of "a new science, psychoanalysis, a part of psychology." But by his own admission, Freud denied his science the one thing it needs to be the *physical* science he always envisioned: "The analyst is unlike other scientific workers in this one respect, that he has to do without the help which experiment can bring to research."

In this regard, in fact, psychoanalysis was no different from the science that Einstein had inadvertently founded, cosmology. Once astronomers mostly agreed on an interpretation of Edwin Hubble's 1929 discovery of a relationship in galaxies between distance and velocity—that the universe is expanding—the discussion quickly faded into vagueness. Some theoreticians took the inevitable step of wondering what such a universe might be expanding *from*. Over the next couple of decades a few theorists hypothesized that the universe existed in a steady state and that fresh matter was somehow emerging from the centers of galaxies. A few others hypothesized that the universe emerged from a single event, called (by a steady stater, ironically) a big bang. But the general theory of relativity itself had exhausted its practical applications regarding further revelations about the universe—an exhaustion that Einstein lived long enough to see. Not that this development would have disappointed Einstein; he always maintained that the importance of the theory was its mathematical elegance, its formal simplicity, not the manifestations of a few nearly undetectable phenomena in the heavens. Still, it's some indication of the impasse cosmology had reached that in a collection of Einstein's writings appearing in 1954, only a year before his death, an editor's note summarized the state of the science, almost four decades after Einstein's first paper on the subject, with this faint praise: "Cosmology is still actively pursued by many scientists. . . ." Or as the astrophysicist Stephen Hawking would one day observe, "Cosmology used to be considered a pseudoscience and the preserve of physicists who might have done useful work in their earlier years, but who had gone mystic in their dotage."

But then cosmology got lucky.

It happened in 1964, when two Bell Telephone Laboratories engineers in New Jersey who were testing a new twenty-foot radio antenna for use in satellite communications found they were picking up a faint signal that they couldn't seem to eliminate from their equipment. They even tried scrubbing the pigeon droppings out of the mouth of the antenna, but the signal persisted. Word of their predicament reached nearby Princeton University, where a group of physicists realized that the wavelength of this radio signal corresponded to a temperature of about 3 degrees above absolute zero (the temperature of -273 degrees Celsius at which all thermal movement of atoms and molecules stops). For these physicists, a 3-degree signal wasn't a nuisance but a number that already had fallen out of the big bang theory: the precise level of energy that—after billions of years of cooling from the highest end of the electromagnetic spectrum to, now, the lowest—should still be resonating throughout the universe.

The discovery of a 3-degree background "noise" didn't exactly confirm the big bang theory, because according to the principles of twentieth-century, logical-positivist science, theories of this kind *can't* be confirmed. Like Newton's theory of gravitation, they can only be validated again and again, to greater and greater levels of precision, until the day they fail, at which point they must be tossed out or amended. The 3-degree background didn't even necessarily falsify the steady state theory, the reigning alternative to the big bang. What this discovery did instead was provide such an exquisite match of mathematical prediction and empirical evidence that it persuaded virtually an entire generation of astrophysicists not only to focus its attention on the big bang but, both human nature and the nature of scientific discovery being what they are, to begin to believe in it.

That nonoptical wavelengths would have any utility for astronomy had been far from a foregone conclusion. Why would they? In a 1923 article in *Scribner's Magazine* called "Seeing the Invisible," for instance, the physicist Robert A. Millikan had introduced a general audience to the idea that other parts of the electromagnetic spectrum are present beyond what our eyes can see, and he had even briefly entertained the notion that one day these other wavelengths might have some astronomical applications. But what he reported about the radio part of the spectrum in particular he also found to be true of all the others: "It has not opened up new worlds to our perceptions."

Now it had. What was lucky for cosmology about the astronomical detection of invisible-light wavelengths wasn't simply that astronomers had found them. The 1964 discovery by Bell Labs engineers was indeed serendipitous. For that matter, so was the first detection of astronomical data in invisible-light wavelengths, in 1932, also at Bell Labs in New Jersey, also by an engineer picking up an unwanted radio signal. (It turned out to be coming from the stars—an odd observation, and one that received some publicity, though mostly for its curiosity value.) But the Princeton physicists had recognized the value of the 3-degree signal because they knew what to look for—were, in fact, already looking for it, planning to build an antenna of their own for the express purpose of picking up a 3-degree signal. What was lucky for cosmology about invisible-light wavelengths wasn't that astronomers had found them but that they could carry valuable, even crucial information. And in order to recognize and exploit their luck, astronomers once again had to learn to adjust to the way they think about the way we see. They had to ask themselves: If radio waves contain valuable information about the heavens, then what about all the other nonoptical wavelengths—infrared, ultraviolet, X-rays and gamma rays? Do they have anything important to tell us? The visible part of the electromagnetic spectrum covers only those wavelengths from 1/700,000 centimeter (red) to 1/400,000 centimeter (violet). On each side of that narrow window of sight, the lengths of the electromagnetic waves lengthen and shorten by a factor of about a million billion, from the longest radio waves to the shortest gamma rays. From the point of view of an astronomer in the middle of the twentieth century, that was a lot of potential information, and certainly worth a look—or "look."

In X-rays, for instance, the universe changed overnight, literally. One minute before midnight on June 18, 1962, in the New Mexico desert, an Aerobee rocket outfitted with several X-ray detectors went up. Then, gravity being what it is—or, at any rate, gravity being *whatever* it is—the rocket came down, though not before reaching a maximum altitude of 140 miles, or 225 kilometers, and remaining above the crucial altitude of 50 miles, or 80 kilometers, for exactly five minutes, fifty seconds. (The shorter, more energetic wavelengths can't penetrate the Earth's atmosphere—and a good thing, too, or our species would have never evolved.) For more than a decade, astronomers had been studying X-rays from the sun, which are a million times fainter than the radiation the sun emits at wavelengths we can see with our eyes. What the Aerobee team had designed this rocket to do was detect X-rays from the moon, which astronomers anticipated would be at least a thousand times fainter still—if those X-rays even exist. And in fact the astronomers who examined the data from the rocket found they *don't* exist; no X-rays emanated from the spot in the sky where the moon was. The astronomers did, however, glean two other results, and they couldn't decide which was the more baffling: the presence of a specific, individual X-ray source radiating with an intensity 100,000,000 times greater than anyone had expected; or the presence of a nonspecific, or diffuse, background that apparently was lighting up the night sky with X-rays as brilliantly as visible light irradiates the day sky at one minute before *noon*.

The universe, apparently, was not what it seemed. If X-rays mean a lot of energy and a lot of sources of X-rays means a lot of sources of a lot of energy, then a lot of sources of a lot of energy permeating the heavens means . . . what? At first, theorists couldn't imagine. And then they could—but they could imagine it only because the math told them they *should* be able to imagine it. Like an expanding universe or, by extension, the cosmic microwave background that validated the big bang cosmology, the answer was already there, embedded in Einstein's relativity, beggaring belief.

In his 1796 *Exposition of the World System,* Laplace had used Newtonian laws to suggest that if a star were massive enough, its gravitational force would be strong enough to keep its rays of light to itself, preventing them from reaching our eyes; "it is therefore possible," he wrote, "that the largest luminous bodies in the universe may, through this cause, be invisible." Such speculation didn't begin to assume modern form, however, until January 1916, when the German astronomer Karl Schwarzschild seized upon the weeks-old notion of relativity to produce his paper "On the Gravitational Field of a Point Mass According to the Einsteinian Theory." In the 1930s, the theorists Lev Landau and Subrahmanyan Chandrasekhar independently arrived at the mathematical expression of what the mass for such an object would have to be in order to trigger such a collapse, and then in 1939 two other physicists, J. Robert Oppenheimer and Hartland S. Snyder, mathematically described the process of the star's collapse—one that involved such Einsteinian notions as light deflection and time dilation. As Oppenheimer and Snyder pointed out, for an observer accompanying the collapse of the star, the time of the collapse will be finite, perhaps "on the order of a day"; for an external observer, however, the time for the collapse, due to the stretching of the electromagnetic waves as the rate of the infall of material speeds up and approaches the velocity of light, will seem "infinite." Even Eddington, Einstein's longtime popularizer, had to shake his head: "There should be a law of nature to prevent a star from behaving in this absurd way."

But even these calculations, like Laplace's speculations, would remain in the realm of the theoretical if astronomers didn't have empirical evidence to support them. Thirty years later, after considering the results of the 1962 X-ray rocket, they were beginning to think they did—that a star collapsing in such a fashion could account for the tremendous outpouring of energy that an X-ray source in the sky suggested. By 1967—the same year that John Archibald Wheeler gave this phenomenon the gift of an unforgettable name, "black hole"—astronomers had catalogued about 30 such candidates; after the launch of the Uhuru satellite in 1970, about 150 sources, including one, called Cygnus X-1, that seemed particularly promising. This was a two-star, or binary, system, one star of which astronomers could observe in visible light behaving in such an erratic fashion as to suggest, by the laws of gravity, that it had an invisible companion. By definition, the invisible companion wouldn't be giving off light, but the gas from the visible star, plunging into the tremendous gravitational field of the collapsing object, would heat up to a billion degrees or more—hot enough that *it* would give off invisible light in the form of X-rays. By 1994, astronomers were willing to commit themselves—to transfer the concept of black hole "from the realm of theory to reality," as *The New York Times* reported as the lead news on May 26 of that year, under a headline that read, in part, "AS EINSTEIN PREDICTED."

"AS FREUD PREDICTED" is not a headline likely to ever appear on any front page. The ability of mathematics to make predictions is not only one that any science might envy but one that Freud wished for psychoanalysis. It's why he tried to trace the paths of nervous energy in his "Psychology for Neurologists," it's why he tried to trace the paths of psychic energy in "The Interpretation of Dreams," and it's why, in the work he completed during his final year, *An Outline of Psychoanalysis,* he was still harkening back to the physical-and-chemical dream of Brücke and Du Bois-Reymond: "the phenomena with which we were dealing do not belong to psychology alone; they have an organic and biological side as well." Or, as he wrote on another occasion regarding the psychoanalytic vocabulary: "The deficiencies in our description would probably vanish if we were already in a position to replace the psychological terms by physiological or chemical ones."

To some extent the "deficiencies" of psychoanalysis relative to a science such as physics are due to a difference in maturity. If we take Freud at his word that he was attempting to found a new science, and that the primary purpose of his decades-long collection of clinical data was to begin to build a theory, then what Einstein said about physics in the seventeenth century would apply to psychoanalysis in the twentieth century: that it relies upon "the predominantly inductive methods appropriate to the youth of science."

In that case, the quality of the data itself must come into question. "The teachings of psychoanalysis are based on an incalculable number of observations and experiences," Freud wrote in the preface to the *Outline,* "and only someone who has repeated those observations on himself and on others is in a position to arrive at a judgment of his own upon it." Freud's purpose in issuing this caveat no doubt was to preempt potential critics. But if the analytic hour is the instrument for discovering and investigating the data of this new science, then case histories can't help but serve as the equivalent of the earliest astronomical or anatomical drawings. They're only as good as the artist. They may be reasonably accurate, but compared with photographs taken with a telescope or microscope (or data stored digitally), they can seem hopelessly subjective, if not primitive, by the standards of modern science.

Still, it was by these standards that the battle over psychoanalysis's legitimacy was fought during Freud's lifetime and, due to the lingering effect of his own insistence, for some time thereafter. As late as 1949—ten years after Freud's death—an American psychoanalyst could still confidently refer to "that steadily decreasing but still substantial group of psychologists for whom both the method and the subject matter of psychoanalysis are outside the realm of natural science." That same year, however, the organizing committee for an influential lecture series at the California Institute of Technology was voting to convene a symposium on psychoanalysis specifically because of "the very fact that there was so much question in the minds of many regarding the scientific status of psychoanalysis."

Like Galileo claiming that he'd seen everything in the heavens that was worth seeing, Freud had overreached. Freud, in fact, tended to extremes, and as his letters and other private papers became public over the course of the 1950s, the conduct of the big, strong man who had led the world by the hand into the darkest reaches of the psyche began to seem more than a little unheroic.

He had oversold psychoanalysis as a therapy. The talking cure was not a "cure" in the sense that it always worked. Even in the case studies that Freud had supplied, the patient's cathartic release and freedom from symptoms, on closer inspection of the historical records, hadn't always lasted. What Darwin had once confessed to a friend in a letter applied to Freud as well: "What you hint at generally is very very true, that my work will be grievously hypothetical & large parts by no means worthy of being called inductive; my commonest error being probably induction from too few facts."

Freud had also oversold psychoanalysis as an instrument of investigation into the mind. By his own admission, Freud had twice encountered obstacles that called the efficacy of the technique into question. The first came in the late 1890s during his attempt to formulate a seduction theory—that adult reports of childhood sexual abuse are all based on actual incidents. But if that were the case, Freud came to realize, then incest and pedophilia had reached epidemic proportions, at least among the cultured classes in Vienna. Freud eventually abandoned the seduction theory and replaced it with an attempt to understand the role of fantasy in adult memories. The second obstacle, perhaps not unrelated, came in his recognition of transference and counter-transference—the ways that the therapist can influence the response of a patient, and vice versa. In both cases, Freud claimed to have transformed these seeming obstacles into virtues: that fantasies, whether based in reality or not, nonetheless reveal a psychic truth; and that in the interaction with the analyst a patient is reliving a relationship with a formative figure from the past—a parent, perhaps. And maybe he had. Certainly psychoanalysts who rely on the principles of fantasy and transference every day, who "see" them in action, would say so. But precisely because of these complications psychoanalysis was not and never could be quite the precise surgical tool, the laserlike scientific instrument, that Freud said he'd discovered.

Freud was always overreaching, overreacting, overselling. He overreached when he appeared before the Imperial Society of Physicians back in 1886 to try to say what he'd learned in Paris, and he overreacted when he later recalled both that his peers had rejected the existence of male hysteria and that he had resigned from the organization (neither was true). He claimed not to care about posterity, going so far as to burn his papers specifically in order to confound his biographers, but wondered whether a plaque might one day mark the spot where he first conceived of the importance of dreams. He repeatedly did to his friends what he'd once done to his father—idolizing Breuer, then turning on him; idolizing Fliess, then turning on him. And in so doing, he left the world ample ammunition to do the same to him—to idolize him while he was alive, to turn on him when he was dead.

"I consider it a great misfortune," he once wrote to Martha, "that nature has not granted me that indefinite something which attracts people." Freud could be his own worst enemy, and time and again the temptation among his critics to use his own surgical instrument to cut him down to size has proved irresistible: to point out that the father of psychoanalysis was hypochondriacal, grandiose, defensive, petty, delusional, wrong. But such exercises themselves get the psychoanalytic process wrong—the very process that they're supposedly wielding with heavy irony. After all, the point of psychoanalytic insight is to say not "Aha!" but "Ah"—not to expose failing but to understand conflicts.

And this is the path that psychoanalysis followed in the second half of the twentieth century: to continue to formulate a theory of the mind that encompasses the unconscious. Yes, Freud oversold psychoanalysis as a theory, too, especially as he grew older and his speculations grew bolder. As he himself wrote in 1920 in his *Jenseits des Lustprinzips* (*Beyond the Pleasure Principle*): "At all events, there is no way of working out this idea except by combining facts with pure imagination many times in succession and thereby departing far from observation. We know that the final result becomes the more untrustworthy the oftener one does this in the course of building up a theory, but the precise degree of uncertainty is not ascertainable. One may have gone ignominiously astray."

Did Freud go astray, and if so how far? As the century progressed, and Freud's immediate influence waned, it has become increasingly possible for psychoanalysts to ask themselves these questions—and in so doing to reclaim psychoanalysis as their own.

In this regard, the contrast with cosmology can also help to highlight a difference not just in maturity but in kind. To ask psychoanalysis to be scientific in the falsifiable sense of the word is either to insist on the impossible or to descend into semantic wrangling over the meaning of science. We can expand the definition of science to include those that aren't strictly falsifiable, which is the path that psychoanalysis, following Freud's lead, chose for the first half of the twentieth century, until falsifiability as the gold standard of science overtook it. Or we can subject these sciences, "sciences" or pseudosciences to their own (presumably high and demanding) standards that have nothing to do with falsifiability.

To some extent the hundredth anniversary of the publication of Darwin's *On the Origin of Species* in 1959 helped reframe the argument. "The most important lesson to be learned from evolutionary theory today," the lead article in an issue of *Science* that year noted, "is a negative one: the theory shows us what scientific explanations need not do or be"—in a word, predictive. What Freud's teacher Carl Claus—the Darwin expert who headed the zoology department when Freud entered the University of Vienna—once said of Darwin still applied, only now it applied to psychoanalysis as well: "It is soon apparent that direct proof by investigation is now, and perhaps always will be, impossible."

The problem was the number of variables. A science like physics could afford to make predictions because it involved only a couple of variables existing in pristine, idealized states. Put two planets in motion around each other and you can describe with mathematical precision their motions unto eternity. Put the universe in motion and, if you were Laplace, you could convince yourself that you could describe with mathematical precision their motions unto eternity, as long as you knew the present positions and velocities of everything (which you didn't, as Laplace knew, and which you couldn't, as Heisenberg would come to fully appreciate after his momentous walk with Einstein in 1926, but that's another story—the story of uncertainty and quantum mechanics).

But put a *person* in motion, and all bets were off.

This insight was a variation on one that Freud already had, though he didn't quite apprehend its impact on the scientific value of his work as a whole. We are overdetermined—the word he coined to describe data that had multiple variables leading to its creation and therefore multiple interpretations as to the meaning of how it got there. The dream of his father, for instance: What did "close the eyes" really mean? Did it refer to the dead man's eternal rest? To a son's desire to avert his eyes? Or did it perhaps refer to psychoanalysis itself—the method Freud was in the middle of developing at that very moment, a method that not only literally required patients to lie still and close their eyes but metaphorically required Freud to close *his* eyes, too, and to rely on data that lay beyond the evidence of the senses?

For three hundred years the number of objects in the heavens and the human body had increased with every improvement to the technology of the telescope and the microscope. Those perceptual advances in turn had fostered an unthinking assumption that the route to further discoveries would be through further technological improvements—and it was indeed *a* route. But the revelations available through nonoptical astronomy or psychoanalytic theory haven't been a matter of seeing more of what was already out there—haven't been a matter of seeing farther or deeper. Like the conceptual leaps that Einstein and Freud made in arriving at relative time or the dynamic unconscious, these perceptual leaps were a matter of finding something *other*—something parallel: a universe that coexists with the one we thought we knew.

Black holes might be extreme cases in astrophysics, but that's precisely what high-energy electromagnetic waves allow astrophysicists to detect: some of the most extreme, energetic, violent phenomena in the known universe. In the early twenty-first century the launch of the Chandra X-ray telescope solved the mysteries lingering from that first rocket launch in 1962. According to the best theories of the day, the individual sources of X-rays have turned out to be black holes within our galaxy; the diffuse X-ray background has turned out to be black holes in other galaxies.

Just as the invisible part of the electromagnetic spectrum overwhelms the sliver of visible, so the unconscious part of our existence overwhelms the conscious. Although it's some indication of the early stage of psychoanalytic theory that metaphors still do more justice to it than empirical evidence, it's also true that the old metaphors need updating: The unconscious isn't the underwater part of an iceberg, and it's not even a submerged continent; it's the ocean itself, deep, vast and, only a hundred years into this new science, still unfathomable.

Over the course of the seventeenth century, the universe was reborn in our imaginations. Not only did Earth come to number among the planets and man among the animals, and not only had Newton and Descartes endeavored to make the motions of Earth and man predictable, but as a result the universe became one that unmistakably moved. Over the course of the twentieth century, we have similarly reimagined the universe. What access to the invisible has allowed us to see is the universe in motion *over time.*

Which was there all along. After all, motion *implies* the passage of time; you can't get from here to there without it.

Einstein took what was inherent in the role that the finite velocity of light played in the Newtonian conception of the outer universe and made it essential. Freud took the subtext in the work of artists and philosophers regarding our inner universe and elevated it to the level of text. So the invisible has allowed us all to take what was implicit in the conception of a universe that moves and make it explicit: It moves over time. When we see the outer universe now—photographs, for instance, of radio-wave remnants of the big bang in full color—we see it as a system with a beginning and, maybe, an end. When we see the inner universe now—the behavior of people we know well or even glancingly meet—we see it as a system with a past that informs the present. We have become man with X-ray eyes.

What is gravity? As of the turn of the twenty-first century, physicists estimate that the galaxies in the universe number 125 billion or so, and that each may contain 50 million black holes, bringing the universe's black hole population to well over 5 quintillion. But they still can't say what happens beyond the event horizon, the black hole's ring of no return.

What is consciousness? As of the turn of the twenty-first century, neuroanatomists estimate that the human neocortex contains 100 billion cells, each with 1,000 to 10,000 synapses, allowing it to make roughly 100 trillion connections. But they still can't say exactly what a thought is.

A century is not a very long time—not in terms of the universe, certainly, and not in the terms of our species and not even in terms of science. Invisible-light astronomy and psychoanalytic technique arose out of a common need and have grown to create a common vision: not that there is more to the universe than meets the naked eye, and not that there is more to the universe than meets a naked eye assisted by instruments that extend the sense of sight, but that there is more to the universe *than meets the eye at all.*

Who knew? After all, nonoptical wavelengths or the thoughts of human beings might easily have revealed nothing beyond what we already knew to be there and what we knew to be there might easily have been all there was to know. The eye aided by instruments that see more than the naked eye alone—that see more than telescopes or microscopes—didn't *have* to be inadequate as a means of investigating nature; the invention of instruments that could see what wasn't visible in the outer and inner universes didn't *have* to open a new frontier. But it was; and they did.

And that—after four hundred years of searching for the secrets of the universe in plain sight, and after a century of seeking further secrets in hidden universes, without and within—is a start.

NOTES

The notes in the following pages cover only the facts and quotes that wound up making their way into this book. Consider them the visible evidence of invisible influences. For readers who want to explore the context for the text of this book or learn more about a particular subject, the descriptions at the beginning of each chapter heading below offer both amplifications of citations already in the footnotes and suggestions for further reading from the bibliography.

EPIGRAPH

p. xi Nitske, p. 131.

PROLOGUE

1 **They met only once:** Freud, 1953–1966, v. XXII, p. 198.
1 **During the New Year's:** Jones, 1953–1957, v. 3, p. 131.
1 **"I regret":** Einstein, 1979, p. 35.
1 **"He understands":** Jones, 1953–1957, v. 3, p. 131.
2 **In 1610:** Galilei, pp. 47 ff.

2 **Beginning in 1674:** *Dictionary of Scientists,* p. 329.
2 **"upwards of one million":** Ibid.
2 **"animals" numbering:** Williams, v. 2.
2 **"an unbelievably":** "Antony van Leeuwenhoek (1632– 1723)," www.ucmp.berkeley.edu/history/leeuwenhoek.html.

ONE: MORE THINGS IN HEAVEN

Brian, Clark and Fölsing are the standard overall biographies of Einstein; see Frank 1989 for a more philosophical focus, and Pais 1982 for a more scientific one. Holton collectively identifies in Einstein's methodology a new "theory of knowledge;" see also his 1995, 1998a, 1998b. Miller 1997 offers a nearly word-by-word examination—as well as a new translation—of Einstein's paper on special relativity; see also his 1986. In addition to Schilpp, anthologies on Einstein's life and work include French, Woolf. For more on Michelson, see Livingston, Michelson.

9 **His father had:** Einstein, 1949, p. 9.
10 **"all but":** Poincaré, 1952, p. 170.
10 **Einstein himself:** Einstein, 1987, pp. 4–5.
11 **"From what":** Einstein, 1987, p. 169.
11 **He'd first thought:** Thompson, p. 1015.
11 **a three-day celebration:** Gray, June 18, 1896, pp. 151–52; Gray, June 25, 1896, pp. 173–75.
11 **"I have not had":** Thompson, p. 1065.
11 **"I never satisfy":** Ibid., p. 835.
12 **In one such demonstration:** Gray, 1908, pp. 290–91.
13 **The lecture hall:** Ibid., p. 279.
13 **George Gabriel Stokes:** Zajonc, p. 121.
13 **Wheatstone wave machine:** Exhibit at the Science Museum, London.
13 **On other occasions:** Gray, 1908, p. 257; Thompson, p. 818.
14 **"The application":** Gray, 1908, p. 258.

14 **His idea was:** Swenson, p. 26.
15 **The Berlin reading:** Fölsing, p. 160.
15 **"One thing we":** Thompson, p. 1035.
15 **"It is to be hoped":** Kelvin, p. vi.
15 **a reading 10 percent:** Swenson, p. 28.
16 **"that the luminiferous":** Ibid.
16 **"I cannot see":** Kelvin, p. 485.
16 **Kelvin was referring:** Pais, 1982, pp. 122–25.
16 **"Thus," Lorentz concluded:** Lorentz, p. 5.
16 **"An explanation was":** Poincaré, 1952, p. 172.
17 **"Surely this course":** Holton, 1988, p. 323.
17 **"I predicted that":** Stachel, 1987, p. 46.
17 **a passing reference:** Pais, 1982, pp. 116–17.
17 **"a stepmotherly fashion":** Stachel, 1987, p. 45.
17 **"a very much simpler":** Fölsing, p. 157.
17 **"If only relentless":** Stachel, 1987, p. 46.
18 **"The introduction of":** Stachel, 1987, p. 45.
19 **"four wandering stars":** Galilei, p. 28.
19 **"a round shape":** Van Helden, 1989, p. 107.
19 **"a semicircle":** Ibid.
19 **"sickle-shaped":** Ibid., p. 108.
20 **"Venus revolves":** Galilei, p. 94.
22 **his school in Aarau:** Einstein, 1949, p. 53; Pais, 1982, p. 131.
23 **"battle":** Pais, 1982, p. 139.
23 **"Thank you!"** Ibid.
23 **"If, for example":** Stachel, 1998, p. 125.
24 **In 1676:** Cohen, p. 346.
25 **"I have looked":** Hoskin, p. 2.
28 **"exercised a profound":** Einstein, 1949, p. 21.
29 **"Absolute motion":** Newton, 1952, p. 9.
29 **"Place is a part":** Ibid.
29 **"Absolute space":** Ibid., p. 8
29 **"It is indeed a matter":** Ibid., p. 12.

29 **"We join with":** Mach, p. 305.
30 **"historically first":** Mach, p. 595.
30 **"shook . . . dogmatic faith":** Einstein, 1949, p. 21.
30 **"We shall raise":** Stachel, 1998, p. 124.
30 **"independent of the state":** Ibid.
30 **"only apparently":** Miller, 1997, p. 371.
30 **the "introduction of":** Stachel, 1998, p. 124.
32 **"inseparable connection":** Pais, 1982, p. 139.
33 **"it supplies us":** Einstein, 1995, pp. 16–17.
34 **"For the rest of":** Clark, p. 252.

TWO: MORE THINGS ON EARTH

Gay and Jones 1953–1957 are the standard overall biographies of Freud; but see also Wollheim 1990. For anthologies on Freud's life and work, see Mujeeb-ur-Rahman, Nelson, Neu, Wollheim 1974. For more on Freudian theory overall, see Waelder. For more on Freud's neurological origins, see Amacher, Erikson, Rapaport and Gill, Sacks. For more on Freud's invocation of the unconscious, see Eidelberg, Giora, Herzog.

35 **His father had:** Freud, 1953–1966, v. IV, p. 197.
36 **"the very threshold":** Williams, v. 4.
37 **"I know that the existing":** Bernfeld, 1949, p. 179.
38 **"The endless confusion":** Clarke and Jacyna, p. 10.
38 **Joseph Jackson Lister:** Jane Insley, "Lister's 1826 Microscope," www.fathom.com/feature/122351.
38 **In 1827:** Clarke and Jacyna, p. 60.
39 **"one sees instead":** Ibid., p. 389.
39 **"However," as one:** Ibid., p. 84.
39 **Joseph von Gerlach:** Williams, v. 4.
40 **But if the cells:** Clarke and Jacyna, p. 99.
40 **Camillo Golgi:** Williams, v. 4.
41 **Ramón y Cajal:** Ibid.

41 **"A connection with":** Clarke and Jacyna, p. 99.
41 **"I am so deep":** Freud, 1954, p. 118; Masson, p. 127.
42 **"congresses":** Masson, p. 2.
42 **Freud began composing:** Ibid., p. 139.
42 **"I am writing":** Ibid.
42 **"The project":** Freud, 1954, p. 355.
43 **"the knowledge of":** Ibid., p. 358.
43 **"I have been":** Freud, 1954, p. 126; Masson, p. 141.
43 **"During one industrious":** Masson, p. 146.
44 **"it required a lot":** Freud, 1954, p. 133; Masson, p. 150.
44 **"I may be able":** Freud, 1954, p. 133.
44 **"I no longer understand":** Masson, p. 152.
45 **Descartes first heard:** Descartes, 1996, p. xx.
45 **has been cultivated:** Descartes, 1985, pp. 114–15.
45 **"there is still":** Ibid., p. 115.
45 **"for the other sciences":** Ibid.
45 **"to demolish everything":** Descartes, 1996, p. 12.
45 **"no knowledge":** Descartes, 1985, p. 115.
46 **He completed both:** Ibid., p. 79.
46 **"mangled":** Ibid.
47 **Richard Mead:** Serota, p. 227.
47 **Franz Anton Mesmer:** Ibid.
47 **Friedrich Herbart:** Margetts, pp. 127–28.
47 **"Regular order":** Jones, 1953–1957, v. 1, p. 373.
48 **Sigmund Exner:** Ibid., p. 380.
48 **He'd used the still-new:** Bernfeld, 1951, p. 214.
48 **He'd twice developed:** Bernfeld, 1949, pp. 181–82.
48 **he'd wondered if cells:** Ibid., p. 179.
49 **"The starting-point":** Freud, 1953–1966, v. XXIII, p. 157.
49 **"A conception":** Ibid., v. XII, p. 260.
49 **"quantity" of:** Freud, 1954, p. 355.
49 **"our most personal daily":** Freud, 1953–1966, v. XIV, p. 166.
49 **"In the interval":** Ibid., v. XIX, p. 14.

50 **Freud had first heard:** Jones, 1953–1957, v. 1, p. 226.
51 **"could describe":** Freud, 1953–1966. v. XX, p. 19.
51 **a "catharsis":** Jones, 1953–1957, v. 1, p. 224.
52 **"The moment at which":** Freud, 1953–1966, v. III, p. 35.
52 **Anna O, for instance:** Ibid., v. II, pp. 38–39.
52 **Frau Cäcilie M:** Ibid., p. 180.
52 **"*Cessante causa*":** Ibid., v. III, p. 185.
53 **That night, in fact:** Masson, p. 202.
53 **"wink at" or "overlook":** Freud, 1953–1966, v. IV, p. 318.
53 **"the most important":** Ibid., p. xxvi.
53 **"By one of those dark":** Masson, p. 202.

THREE: GOING TO EXTREMES

Nitske is one biography of Röntgen; see also Glasser. For more on science at the turn of the twentieth century, see Bernhard, et al., Bunge and Shea, Heilbron.

55 **On November 8:** Nitske, pp. 3–5.
56 **"*Der Röntgen ist*":** Ibid., p. 100.
56 **On December 22:** Ibid., p. 5.
56 **On New Year's Day:** Ibid., pp. 98–99.
56 **The first public:** Ibid., pp. 112–13.
57 **"Men of science":** "Hidden Solids Revealed," *The New York Times,* Jan. 16, 1896.
57 **when an English:** "Prof. Roentgen's X-Rays," *The New York Times,* Feb. 5, 1896.
57 **By January 13:** Nitske, pp. 101–102.
57 **Röntgen granted only:** Ibid., p. 126.
57 **"In a few days":** Ibid., p. 100.
57 **a "longitudinal vibration":** "No New Light Found," *The New York Times,* Feb. 9, 1896.
58 ***Science* magazine reported:** Nitske, p. 121.

58 **A Mr. Ingles Rogers:** Ibid.

58 **a Dr. Baraduc:** "A Test of Credulity," *The New York Times,* June 28, 1896.

58 **"go inside the house":** Edwin E. Slosson, "Sun Dogs," *The Independent,* Oct. 2, 1920, p. 10.

58 **though he did tentatively:** Nitske, p. 317.

58 **"no hospital in the land":** "About X-Ray Photography," *The New York Times,* Sept. 6, 1896.

60 **"I am confident":** Descartes, 1993.

60 **"It was granted":** Kolb, p. 80.

60 **"All celestial mysteries":** Van Helden, 1974, p. 54.

61 **Newton wrote in a letter:** Zajonc, p. 85.

61 **He was paraphrasing:** Ibid.

61 **"We are like dwarfs":** Ibid., p. 79.

63 **In 1655:** Learner, p. 20.

63 **In 1671 and 1672:** North, p. 345.

63 **then, in 1675:** Ibid., p. 344.

63 **"the naked eye":** *Oxford English Dictionary* online.

64 **"No one would claim":** Badash, p. 52.

64 **"It seemed as though":** Ibid., p. 55.

64 **"sixth place of":** Weiner, p. 9.

65 **"all that was left":** Badash, p. 55.

65 **"If the public":** Ibid., p. 53.

65 **"that the opinion":** Ibid., p. 50.

66 **"Accurate and minute":** Ibid., p. 57.

66 **"A hundred years ago":** Swenson, pp. 27–28.

68 **"The step":** Pais, 1982, p. 163.

68 **"It is not improbable":** Jo Baer, "Mach Bands," www.ubu.com/aspen/aspen8/machBands.html.

68 **One person, in fact:** Miller, 1984, p. 118.

69 **Many readers simply:** Ibid.

69 **"superfluous":** Stachel, 1998, p. 124.

70 **In 1893:** Galison, p. 228.

70 **in 1902:** Ibid.
70 **"It is clear":** Poincaré, 1982, pp. 233–34.
71 **"The clocks synchronized":** Fölsing, p. 163.
71 **"It is so exceeding":** Cohen, p. 346.
71 **A train not following:** Galison, p. 222.
72 **"master not only":** Ibid., p. 231.
72 **As part of Einstein's work:** Ibid., p. 220–21.
72 **Einstein moved his family:** Ibid., p. 225.
72 **"An analysis":** Ibid.
72 **"We must, for example":** Poincaré, 1982, p. 143.
72 **"unrecognizably was anchored":** Einstein, 1949, p. 53.
72 **"The overwhelming majority":** Freud, 1953–1966, v. XIX, p. 216.
73 **(and Freud himself admitted:** Ibid., v. XIV, p. 15.
73 **Galen had written:** Whyte, p. 78.
73 **St. Augustine asked:** Ellenberger, 1957, p. 3.
73 **"Our clear concepts":** Whyte, p. 99.
73 **Forms of the word:** Ibid., pp. 66–67.
74 **"There is in intelligence":** Ibid., pp. 160–161.
74 **"The key to the":** Ibid., p. 149.
74 **Eduard von Hartmann:** Ibid., pp. 163–65.
74 **By 1890:** James, v. 1, pp. 224–90.
75 **"By Dr. Josef":** Freud, 1953–1966, v. III, p. 26.
75 **Freud, in fact:** Ibid.
75 **"the determining process":** Ibid., v. II, p. 7.
76 **a "hypnoid state":** Ibid., v. I, p. 149; v. II, pp. 12, 215–22; v. III, p. 39.
76 **"I shall call":** Ibid., v. III, p. 47.
76 **"I willingly adhere":** Ibid., v. II, p. 286.
76 **"I find, however":** Ibid., v. III, p. 195.
76 **"This opened the":** Ibid, v. XX, p. 23.

FOUR: A LEAP OF FAITH

For more on the history of mathematics, see Bell 1992, Ekeland 1988, Ekeland 1993.

81 **In his mind:** Pais, 1982, p. 179.
82 **"most fortunate thought":** Ibid., p. 178 (new translation by Barbie Bischoff).
83 **"I must confess to you":** Fölsing, p. 235.
83 **"handsome":** Ibid., p. 102.
83 **"eight hours of fun":** Ibid.
83 **"as the library":** Ibid., pp. 168–69.
83 **"enormously varied":** Ibid., pp. 101–02.
83 **"Many-sided":** Ibid., p. 102.
83 **"When you pick up":** Ibid., p. 104.
83 **"the inventor's way":** Ibid.
84 **Einstein added a section:** Pais, 1982, p. 179.
84 **"I really could have gotten":** Einstein, 1949, p. 15.
85 **At the age of twelve:** Ibid., p. 9.
86 **"deep religiosity":** Ibid., p. 3.
86 **"deeply irreligious":** Ibid.
86 **Albert received his:** Einstein, 1987, pp. xix–xx.
86 **"soon reached the":** Einstein, 1949, p. 5.
87 **acquired all the textbooks:** Einstein, 1987, p. xx.
87 **"The objects with which":** Einstein, 1949, p. 11.
88 **"faith in the existence":** Einstein, 1988, p. 262.
88 **"lived in an age":** Ibid.
88 **"The assumptions made":** Frank, 1949, p. 221.
89 **In 1600 Kepler:** Kolb, p. 55.
89 **But instead of ignoring:** Ibid., p. 64.
89 **Kepler soon determined:** Ibid., p. 61.
90 **"Ah, what a foolish":** Ibid., p. 65.
91 **In 1588, however:** Friedman, p. 30.
91 **After reading William:** Boorstin, p. 311.

92 **apocryphal-sounding but:** Friedman, p. 32.
93 **To test these two ideas:** Weinberg, p. 11.
93 **"pretty nearly":** Weaver, 1957, p. 1229.
93 **This was the question:** Westfall, p. 403.
93 **"the question *how*":** Einstein, 1988, p. 255.
94 **"An intelligence knowing":** Spangenburg and Moser, p. 25.
95 **Laplace himself:** "Pierre-Simon, Marquis de Laplace," *Encyclopædia Britannica* Online, http://search.eb.com/eb/article?eu=48265.
96 **During this period:** Pais, 1982, p. 187.
96 **He submitted a paper:** Ibid., p. 201.
97 **"I would ask":** Ibid., p. 211.
97 **Once, Einstein went:** Fölsing, p. 325.
97 **In Einstein's imagination:** Frank, 1989, pp. 95–96; Bernstein, pp. 95–96, 98–99.
98 **"Grossmann, you've got":** Fölsing, p. 314.
98 **lent Einstein his math:** Pais, 1982, p. 44.
98 **get a job at the Bern:** Ibid., p. 46.
98 **intervened on Einstein's:** Ibid., pp. 208–09.
98 **one of the founders:** Ibid., p. 208.
99 **Back in 1901:** Einstein, 1987, p. 190.
99 **"not obviously permissible":** Pais, 1982, p. 201.
99 **"Compared to this problem":** Fölsing, p. 315.
99 **among the courses:** Bernstein, p. 108.
100 **"Never before in my life":** Overbye, p. 239.
100 **"I now see that one":** Clark, p. 181.
101 **"Nothing more can be done":** Fölsing, p. 319.
101 **On Christmas Eve:** Ibid., p. 306.
101 **In 1840:** Roseveare, p. 20.
102 **By his calculations:** Ibid., p. 11.
103 **hundreds of observations:** Ibid., p. 21.
103 **38 arc seconds:** Ibid., p. 23.
103 **42.95:** Ibid., p. 41.
103 **an estimate of a planet's:** Ibid., p. 23.
103 **christened Vulcan:** Ibid., p. 25.

103 **a belt of asteroids:** Ibid., p. 24.
103 **a moon of Mercury:** Pais, 1982, p. 254.
103 **"material cause":** Roseveare, p. 20.
103 **a few even claimed:** Ibid., p. 25.
103 **In the end, some:** Ibid., pp. 95–113; Pais, 1982, p. 254.
104 **30 arc *minutes*:** Overbye, p. 275.
104 **18 arc seconds:** Ibid.
104 **"The gravitational affair":** Ibid., p. 276.
104 **the following year:** Pais, 1982, pp. 243–44.
104 **A year later:** Ibid., pp. 245, 250.
105 **"completely lost confidence":** Ibid., p. 250.
105 **The following week:** Ibid., pp. 252–53.
105 **43 arc seconds:** Ibid., p 255.
105 **Something burst:** Ibid., p. 253.
105 **"For a few days":** Ibid.
105 **Einstein was talking:** Clark, p. 287.
106 **One reason he'd received:** Ibid.
107 **"Today joyous news":** Fölsing, p. 439.
107 **"I am now completely":** Ibid., p. 321.
107 **the outbreak of war:** Ibid., pp. 356–57.
107 **it still predicted:** Pais, 1982, p. 303.
107 **"These facts must":** Einstein, 1916, p. 145.
108 **"This agreement between":** Roseveare, p. 182.
108 **"REVOLUTION IN SCIENCE":** "Revolution in Science," *The Times of London,* Nov. 7, 1919.
108 **"LIGHTS ALL ASKEW":** "Lights All Askew in the Heavens," *The New York Times,* Nov. 10, 1919.
109 **"one of the most momentous":** "Revolution in Science," *The Times of London.*
109 **a special joint meeting:** Ibid.
109 **"EINSTEIN V. NEWTON":** "The Revolution in Science," *The Times of London,* Nov. 8, 1919.
109 **"This report affords":** Henry Norris Russell, "The Heavens in December, 1919," *Scientific American,* Nov. 29, 1919.

109 **"The world is a curious":** Fölsing, p. 455.
109 **"I feel like a graven":** Ibid., p. 457.
109 **to the isolating effects:** Pais, 1982, p. 306.
109 **a British affirmation:** Ibid., p. 308.
110 **the seeming absurdities:** "Jazz in Scientific World," *The New York Times,* Nov. 16, 1919.
110 **"the most remarkable":** "Revolution in Science," *The Times of London.*
111 **"Cosmological Considerations":** Einstein, 1917, p. 186.
112 **One such mind:** Friedmann, pp. 839–43.
112 **First he found fault:** Bernstein and Feinberg, p. 11.
112 **at least one of the smudges:** Hubble, 1925, pp. 139–42.
113 **and then in 1929:** Hubble, 1929, pp. 168–73.
113 **"gravely detrimental":** Einstein, 1919, p. 193.
113 **"for the purpose of":** Einstein, 1917, p. 187.
114 **In 1931:** Clark, pp. 518–527.
114 **"greatest blunder":** Clark, p. 270.

FIVE: THE DESCENT OF A MAN

For more on Freud and Darwin, see Phillips, Ritvo 1965, Ritvo 1974. For more on Freud and hypnotism, see Blum. For more on the history of the unconscious, see Frey-Rohn, Kelly.

115 **In his mind:** Freud, 1953–1966, v. I, pp. 25–31.
116 **At a society meeting:** Ibid., p. 24; Ellenberger, 1970, p. 441.
116 **During his first appearance:** Ellenberger, 1970, pp. 439–41; Freud, 1953–1966, v. I, pp. 5–15; Sulloway, pp. 38–39.
117 **"I was unable to find":** Sulloway, p. 39.
117 **As was customary:** Ellenberger, 1970, p. 440.
118 **October 13, 1885/February 28:** Sulloway, p. 28.
118 **he was planning/he'd diverted:** Freud, 1953–1966, v. I, pp. 3–4.
118 **Freud had first:** Ibid., v. XX, p. 9.

118 **It was upon:** Ibid., p 8.

119 **"help suffering humanity":** Ibid., p. 253.

119 **"to become a natural scientist":** Gay, p. 24.

119 **"asked what my greatest":** Ibid., p. 26.

120 **In the late 1530s:** Singer, pp. 177–178.

120 **The result, in 1543:** *Dictionary of Scientists,* p. 537.

121 **In the book's first edition:** Singer, p. 178.

121 **"the septum of the heart":** Ibid., p. 179.

121 **"put his own hand":** Ibid., p. 177.

121 **"animal spirits":** Ramon M. Cosenza, "Spirits, Brains and Minds: The Historical Evolution of Concepts of the Mind," www.epub.org.br/cm/nlb/history/mind=history_i.html

121 **"which, as the first":** Descartes, 1993.

121 **"who are not versed":** Ibid.

122 **"from the very arrangement":** Ibid.

122 **"I could set out many":** Descartes, 1985, p. 97.

122 **William Harvey:** *Dictionary of Scientists,* pp. 240–41.

122 **"The nerves of the machine":** Williams, v. 2.

123 **"the innermost part":** Descartes, 1985, p. 340.

123 **Goethe wasn't the author:** Freud, 1953–1966, v. XX, p. 8.

124 **"bringing man and animals":** Clarke and Jacyna, p. 39.

124 **"Man is animal":** Ibid., p. 38.

124 **"At man the series":** Ibid., p. 39.

124 **"I was already alive":** Freud, 1953–1966, v. XXII, p. 173.

124 **"light will be thrown":** Charles Darwin.

124 **keep on his bookshelf:** Ritvo, 1990, p. 65.

124 **"Man is not a being":** Freud, 1953–1966, v. XVII, p. 141.

124 **"I see no possible":** Erikson, p. 42.

125 **"the theories of Darwin":** Freud, 1953–1966, v. XX, p. 8.

125 **Carl Claus:** Ritvo, 1990, pp. 113–14.

125 **Freud spent his first:** Ibid., p. 114.

125 **in the second semester:** Ibid., p. 115.

125 **In his fourth semester:** Ibid.

125 **The following year:** Ibid., p. 116.

125 **in March 1877:** Ibid.
125 **Arisotle, who had himself:** Bernfeld, 1949, p. 165.
126 **By this time:** Ritvo, 1990, p. 116.
126 **Thirty years earlier:** Cranefield, p. 408.
126 **"an imponderable agent":** Clarke and Jacyna, p. 194.
126 **they had a club:** Bernfeld, 1944, p. 349.
126 **This was the foundation:** Ibid.
127 **To Freud fell:** Bernfeld, 1949, p. 176.
127 **From there Freud:** Ibid., p. 178.
127 **"The subject which":** Freud, 1953–1966, v. XX, p. 10.
128 **"the *bête noire*":** Ibid., v. I, p. 41.
128 **"general nervousness":** Ibid., pp. 41–42.
128 **Napoleon of Neuroses:** Ellenberger, 1970, p. 95.
128 **"when the Lord":** Freud, 1959, p. 11.
128 **a print that hung:** Freud, 1953–1966, v. III, p. 18.
128 **not of priests:** Ibid., p. 20.
128 **Charcot transformed:** Sulloway, p. 29.
129 **Charcot instituted:** Ellenberger, 1970, pp. 90, 93.
129 **"young" science:** Freud, 1953–1966, v. III, p. 11.
129 **"really shameful":** Gay, p. 48.
129 **"seem to me of a different":** Freud, 1959, p. 187.
129 **"cheat one":** Jones, 1953–1957, v. 1, p. 184.
129 **"all you hear":** Freud, 1959, p. 182.
129 **"As you realize":** Ibid., p. 88.
129 **"not at all adapted":** Freud, 1953–1966, v. I, p. 8.
129 **did manage to perform:** Ibid.
130 **"I sometimes come out":** Freud, 1959, p. 185.
130 **"the work of anatomy":** Freud, 1953–1966, v. I, p. 11.
130 **"no more than an expression":** Ibid.
130 **Each Monday:** Ibid., p. 9.
130 **On Tuesday he conducted:** Ibid.
130 **The rest of the week:** Ibid., pp. 9–10.
131 **produce a patient:** Ellenberger, 1970, p. 98.
131 **mimic symptoms:** Ibid., p. 96.

131 **a recent invention:** Ibid.
131 **highlighting the tremors:** Ibid.
131 **Most dramatic of all:** Ibid., p. 95.
131 **"Paris is simply":** Freud, 1959, p. 188.
131 **Overhearing Charcot:** Freud, 1953–1966, v. XX, p. 12.
131 **a frequent guest:** Jones, 1953–1957, v. 1, pp. 186–87.
132 **"What a magic city":** Freud, 1959, p. 206.
133 **Four years earlier:** Jones, 1953–1957, v. 1, p. 103.
133 **counseled him for financial:** Freud, 1953–1966, v. XX, p. 10.
133 **he need never see patients:** Ibid., p. 253.
133 **soon received the title:** Bernfeld, 1951, p. 209
134 **soon gravitated to the Laboratory:** Ibid., p. 210.
134 **There he distinguished himself:** Ibid., pp. 211–13.
134 **the effects of cocaine:** Freud, 1953–1966, v. XX, pp. 14–15.
134 **To his horror:** Ibid., p. 15.
134 **Immediately upon his return**. Sulloway, p. 35.
134 **He also sent out notices:** Gay, p. 53.
134 **begin planning his wedding:** Sulloway, p. 35.
135 **"provided detailed instructions":** Freud, 1953–1966, v. XX, p. 16.
135 **only because a handful:** Bernfeld, 1951, p. 209.
135 **"The realization":** Freud, 1953–1966, v. XX, p. 16.
136 **Freud found somewhat:** Ibid., v. II, p. xi.
136 **his first child:** Jones, 1953–1957, v. 1, p. 152.
136 **decided to try hypnosis:** Freud, 1953–1966, v. XX, p. 16; Masson, p. 17.
136 **Charcot used the technique:** Ellenberger, 1970, p. 91.
136 **who in 1882 had argued:** Ibid., p. 90.
136 **"a human being is":** Sulloway, p. 42.
136 **"a decapitated frog":** Serota, p. 237.
136 **"to experiment on a human":** Freud, 1953–1966, v. I, p. 98.
136 **First Charcot would induce:** Ellenberger, 1970, p. 91.
137 **Freud had included:** Freud, 1953–1966, v. I, p. 13.
137 **twice lectured on the topic:** Jones, 1953–1957, v. 1, p. 229.

137 **he'd contracted to translate:** Masson, p. 17.
137 **only for the money:** Ibid., p. 24.
137 **"We must agree":** Freud, 1953–1966, v. I, p. 81.
138 **"use of hypnotic suggestion":** Ibid., p. 75.
138 **Freud, of course:** Jones, 1953–1957, v. 1, p. 226.
138 **"showed no interest":** Freud, 1953–1966, v. XX, pp. 19–20.
138 **"allowed it to pass":** Ibid., p. 20.
138 **Freud now proved:** Ibid., p. 17.
138 **Freud traveled back:** Ibid.
138 **"But as soon as I":** Ibid., v. II, p. 108.
139 **It came in the autumn:** Ibid., pp. 110, 135.
139 **"lo and behold!"** Ibid., p. 110.
139 **He'd laid his hand:** Ibid., p. 109.
139 **Now Freud tried:** Ibid., pp. 110, 145.
139 **"You will think":** Ibid., p. 110.
140 **Freud had mentioned:** Ibid., v. XX, pp. 13–14.
141 **"the structure of the":** Ibid., v. I, p. 49.
141 **It appeared in July 1893:** Masson, p. 52.
141 **Charcot left Paris:** Ellenberger, 1970, p. 100.
141 **In the copy of the issue:** Masson, p. 23.
141 **"I will attempt":** Freud, 1953–1966, v. I, p. 169.
141 **"will be an alteration":** Ibid., p. 170.
142 **"the chemical physical:"** Bernfeld, 1949, p. 171.
142 **"Hysterics suffer":** Freud, 1953–1966, v. II, p. 7.
143 **"In what follows":** Ibid., p. 185.
143 **Yet it was Breuer:** Ibid., pp. 192 ff.
143 **"it still strikes me":** Ibid., p. 160.
143 **"irrespectively of whether":** Ibid., p. 279.
143 **And then they changed:** Ibid.
144 **"I couldn't believe":** Ibid.
144 **"like an opera prince":** Ibid.
144 **"I could have told you":** Ibid.
144 **"not knowing" suggested:** Ibid., p. 270.
145 **"must no doubt":** Ibid., p. 268.

145 **a distinct advantage:** Ibid., pp. 268–70; v. IV, p. 16; v. XX, pp. 27–29.
146 **"the triumph of my life":** Ibid., v. XX, p. 253.
147 **He was mourning:** Ibid., v. IV, p. xxvi.
147 **A slip of the tongue:** Masson, pp. 326–27.
147 **Freud wrote it up:** Freud, 1953–1966, v. III, pp. 289–97.
148 **"The intention of this":** Freud, 1954, p. 355.
148 **"hypnotic analysis":** Freud, 1953–1966, v. III, p. 59.
148 **"psychical analysis":** Ibid., p. 47:
148 **"psychological analysis":** Ibid., p. 75.
148 **"clinico-psychological analysis":** Ibid., p. 53.
148 **Then on February 5:** Ibid., p. 143.
148 **"owed his results":** Ibid., p. 150.
148 **"laborious but completely":** Ibid., p. 162.

SIX: A DISCOURSE CONCERNING TWO NEW SCIENCES

For more on relativity and turn-of-the-twenty-first-century astrophysics, see Bartusíak, Coles, Turner and Tyson, Will. For more on psychoanalysis and turn-of-the-twenty-first century neurobiology, see Gabbard, Reiser, Schore, Solms. For more on the epistemological implications of Einstein's work, see Blackmore, Bohr 1987a, Bohr 1987b, Bohr 1987c, Heisenberg 1949, Heisenberg 1958, Heisenberg 1974, Planck. For more on the epistemological implications of Freud's work, see Beecher, Bowers and Meichenbaum, Brower, Bruck, Crews 1995, Crews 1998, Dilman, Edelson, Feigl and Scriven, Gabriel, Gill, Grünbaum 1985, Grünbaum 1998, Hook, Jones 1918, Kubie, Lewin, Nicoll, Opatow, Rivers, Roth, Shope, Shuey, Sterba, Wolman. For more on the epistemological considerations of both the hard and soft sciences, see Adrian, Frank 1956, Frank 1957, Hawkins, Weaver 1955.

153 **"What, precisely":** Einstein, 1949, p. 7.
154 **"something like my own":** Ibid., p. 3.

154 **"remains the one light":** Freud, 1953–1966, v. XXIII, p. 286.
154 **"Every science is based":** Ibid., p. 159.
156 **"which have produced":** "Wonderful Things Done By the Camera," *The New York Times,* Sept. 28, 1895.
157 **"unique":** "About X-Ray Photography," *The New York Times.*
157 **On March 1, 1896:** Pais, 1988, p. 42.
157 **"silhouettes":** Ibid., p. 46.
157 **Later that same year:** Ibid., p. 76.
157 **The year after that:** Ibid., p. 85.
157 **And then the year:** Ibid., pp. 53–56.
158 **"In 1894":** Badash, p. 55.
159 **"the Copernican system":** Armitage, p. 194.
159 **"I wish it to be":** Descartes, 1993.
160 **"*Hypotheses non fingo*":** Newton, 1999, p. 943.
160 **"*Cogito ergo sum*":** Descartes, 1993.
160 **"He endures always":** Newton, 1999, p. 941.
160 **"as soon as I supposed":** Descartes, 1993.
160 **"That gravity should be":** Miller, 1984, pp. 71–72.
161 **"the motions of the planets":** Newton, 1999, p. 382.
161 **"If only we could":** Ibid.
162 **"occult":** Dingle, p. 537.
162 **"You have written":** Bell, 1986, p. 181.
162 **"I would give":** Francis Darwin, v. II, p. 7.
162 **"Every astronomer":** Badash, p. 53.
163 **"The Newtonian theory":** Roseveare, p. 95.
163 **Einstein cited:** Einstein, 1949, p. 21.
163 **"A person who knew the world":** Mach, p. 610.
164 **Auguste Comte:** "Positivism," *Encyclopædia Britannica* Online, http://search.eb.com/eb/article?eu=115436.
165 **We can understand any natural:** Frank, 1989, p. 45.
165 **"What are matter and force":** Frank, 1949, p. 93.
165 **"*Ignorabimus*":** Frank, 1989, p. 45.
165 **Olympia Academy:** Clark, pp. 78–81; Fölsing, pp. 99–100.
166 **a 1911 manifesto:** Holton, 1993, pp. 12–14.

166 **"What is the nature":** Miller, 1986, p. xv.
166 **"In your last letter":** Holton, 1988, p. 247.
167 **"A gain in meaning":** Freud, 1953–1966, v. XIV, p. 167.
168 **"I did not think":** Nitske, p. 134.
168 **Once, Einstein met Mach:** Frank, 1989, pp. 104–105.
169 **didn't live long enough:** Holton, 1988, p. 248.
169 **"This theory is not speculative":** Ibid., p. 247.
169 **"Mach's system studies":** Ibid., p. 257.
170 **"The fictitious character":** Einstein, 1988, pp. 273–74.
171 **an intellectual debt:** Einstein, 1949, p. 21.
171 **"By and by I despaired":** Ibid., p. 53.
172 **"Fields," he called:** Zajonc, p. 136.
172 **on April 10, 1846:** Ibid., p. 135.
172 **"This change in the conception":** Einstein, 1988, p. 269.
173 **"The orbits were":** Ibid., p. 265.
173 **"that this speculative":** Holton, 1988, p. 250.
173 **"metaphysical 'original sin.'"** Schilpp, 1949, p. 673.
173 **"I believe," he wrote:** Einstein, 1988, p. 342.
173 **"If you want to find out":** Ibid., p. 270
174 **Werner Heisenberg:** Heisenberg, 1971, pp. 62 ff.
175 **"succession of daringly":** Freud and Ferenczi, p. 55.
175 **"Its aim is to":** Freud, 1953–1966, v. XXII, p. 170.
176 **"was by no means the prevailing":** Einstein, 1988, p. 272.
176 **"all knowledge of reality":** Ibid., p. 271.
176 **"the fundamental concepts":** Ibid., p. 272.
176 **"Experience remains":** Ibid., p. 274.
176 **"The new Columbus":** Frank, 1989, p. 183; Francis D. Murnaghan, "The Quest of the Absolute," *Scientific American,* March 1921.
176 **"not the discovery":** "The Revolution in Science," *The Times of London.*
177 **"I have all the enthusiasm":** J. S. Ames, "Einstein's Law of Gravitation," *Science,* March 12, 1920.
177 **"the inner Africa":** Whyte, p. 132.

177 **"I am nothing":** Masson, p. 398.

178 **"men of science":** "Hidden Solids Revealed," *The New York Times.*

178 **William Herschel:** Lubbock, pp. 262–65.

178 **Hertz used Maxwell's:** Hoffmann, p. 2.

178 **"Darkness Need Not":** "X-Rays Ordinary Light," *The New York Times,* March 22, 1896.

178 **"The great majority":** "The Cathode and X-Rays," *The New York Times,* March 15, 1896.

179 **"radiations are traveling":** "Sun Dogs," *The Independent.*

179 **"We went from the lecture":** Maurice Samuel, "Mr. Einstein Lectures," *Living Age,* April 21, 1921 (reprinted from *The Manchester Guardian*).

179 **"The reader," Eddington:** A. S. Eddington, "Einstein's Theory of Space and Time," *The Contemporary Review,* December 1919.

180 **"It will be seen":** Eddington, p. 32.

180 **"It does not matter":** Edwin E. Slosson, "That Elusive Fourth Dimension," *The Independent,* Dec. 27, 1919.

180 **"There is no illusion":** "Einstein's Theory of Space and Time," *The Contemporary Review.*

180 **"Psychoanalysis," Freud wrote:** Freud, 1953–1966, v. XIV, p. 301.

181 **sunburned flesh; skin stripping:** "About X-Ray Photography," *The New York Times.*

181 **"Why, I have a patient":** "Doctors Discuss the X-Ray," *The New York Times,* Nov. 10, 1903.

181 **"show only slight interest":** Freud, 1953–1966, v. XI, p. 165.

182 **"But creative writers":** Ibid., v. IX, p. 8.

182 **Freud privately complained:** Ibid., v. XX, p. 48; Jones, 1953–1954, v. 2, pp. 6–7; Masson, p. 402..

182 **Wednesday Psychological Society:** Jones, 1953–1957, v. 2, p. 8; Gay, p. 173.

182 **moved his family:** Masson, p. 30.

183 In 1902 Freud received: Ibid., pp. 455–457.
183 "congratulations and flowers": Ibid., p. 457.
183 In 1908 the first: Jones, 1953–1957, v. 2, pp. 38–45.
183 The following year: Ibid., pp. 53–58.
183 In 1905 in *Drei*: Freud, 1953–1966, v. VII, pp. 125–245.
184 "If our common view of nature": "Einstein's Theory of Space and Time, *The Contemporary Review.*
185 "cheerful, full of himself": Brian, p. 157.
185 cancer-afflicted: Ibid.
185 "Here is yet another": Einstein, 1988, p. 232.
186 "relativity theory": Holton, 1982, p. xv.
186 "so-called relativity": Ibid.
186 *Invariantenheorie*: Ibid.
186 "The knowledge that has": Freud, 1953–1966, v. XVIII, pp. 177–78.
186 "No doubt there have been": Ibid., v. XXII, p. 175.
187 "belief and disbelief": Fölsing, p. 651.
187 "beauty and clarity": Ibid.
187 "for many a beautiful": Ibid.
188 "Because man has within him": Freud, 1953–1966, v. XXII, p. 201.
188 "you yourself have said": Freud, 1953–1966, v. XXII, p. 203.
188 "The lucky fellow": Jones, 1953–1957, v. 3, p. 131.
188 "a happy one": Fölsing, p. 609.
188 "Why do you emphasize": Ibid.
188 "could work at mathematical": Jones, 1953–1957, v. 3, p. 154.
188 "They cheer me": Fölsing, p. 457.
189 "third blow to man's": Freud, 1953–1966, v. XVII, p. 143.
189 "that *the ego is not*": Ibid.
189 "No wonder, then": Ibid.
189 "comparable with, or perhaps": "Einstein's Theory of Space and Time," *The Contemporary Review.*
189 the Vienna Circle: "Logical Positivism," *Encyclopædia Britannica* Online, http://search.eb.com/eb/article?eu=49927.

190 **"is a hypothesis":** Freud, 1953–1966, v. VIII, p. 14.

190 **"And here is the opportunity":** Ibid., v. XII, pp. 265–266.

190 **"natural science like":** Ibid., v. XXIII, p. 158.

190 **"to ask us to think":** Field, et al., p. 414.

191 **"The scientific level":** Watson, p. 93.

191 ***doesn't* work and *shouldn't* work:** Popper, 2002, pp. 39, 42.

191 **what he called its falsifiability:** Popper, 1968, pp. 40–41, 78–92.

191 **"figure of speech":** Freud, 1953–1966, v. XVIII, p. 236; v. XX, p. 31.

191 **"a new science":** Audiotape, The Freud Museum, London.

191 **"The analyst is unlike":** Sigmund Freud, "New Introductory Lectures on Psychoanalysis," www.angelfire.com/on/pisd/archive/SigmundFreud.htm.

192 **he always maintained:** Pais, 1982, p. 273.

192 **"Cosmology is still":** Einstein, 1988, p. 218.

192 **"Cosmology used to be":** Hawking and Penrose, p. 75.

193 **It happened in 1964:** Kolb, pp. 238–41, 256–59.

194 **"It has not opened up":** Robert A. Millikan, "Seeing the Invisible," *Scribner's Magazine,* 1923.

195 **One minute before midnight:** Giacconi, et al., p. 64.

195 **For more than a decade:** Lang and Gingerich, p. 62.

195 **What the Aerobee team:** Ibid., pp. 62–63.

196 **two other results:** Ibid., p. 63.

196 **"it is therefore possible":** "D,08 Why can't light escape from a black hole?," www.faqs.org/faqs/astronomy/faq/part4/section-10.html.

196 **January 1916:** Fölsing, p. 384.

196 **"On the Gravitational":** Schwarzschild, pp. 452–55.

196 **In the 1930s:** Lang and Gingerich, p. 456.

197 **then in 1939**. Oppenheimer and Snyder, pp. 455–59.

197 **"There should be a law":** Lang and Gingerich, p. 457.

197 **"black hole":** Wheeler, p. 8.

197 **about 30 such:** Christine Jones, William Forman, and William Liller, "X-Ray Sources and Their Optical Counterparts—I," *Sky and Telescope,* November 1974.

197 **about 150 sources:** Alan P. Lightman, "Some Recent Advances in X-Ray Astronomy," *Sky and Telescope,* October 1976.

198 **a billion degrees:** Ibid.

198 **"from the realm":** John Noble Wilford, "Space Telescope Confirms Theory of Black Holes," *The New York Times,* May 26, 1994.

198 **"the phenomena with which":** Freud, 1953–1966, v. XXIII, p. 195.

198 **"The deficiencies":** Ibid., v. XVIII, p. 60.

199 **"the predominantly inductive":** Einstein, 1988, p. 282.

199 **"The teachings of psychoanalysis":** Freud, 1953–1966, v. XXIII, p. 144.

199 **"that steadily decreasing":** Benjamin, p. 139.

200 **"the very fact":** Hilgard, et al, p. v.

200 **"What you hint at":** Max, p 66.

200 **The first came:** Masson, pp. 264–266.

200 **The second obstacle:** Freud, 1953–1966, v. XX, pp. 42–43.

201 **he later recalled:** Ibid., pp. 15–16.

201 **(neither was true):** Ellenberger, 1970, pp. 437–42; Sulloway, p. 42.

201 **to burn his papers:** Freud, 1960, pp. 140–141.

201 **wondered whether a plaque:** Masson, p. 417.

201 **"I consider it":** Freud, 1960, p. 199.

202 **"At all events":** Frenkel-Brunswick, p. 288.

203 **"The most important":** Scriven, p. 477.

203 **"It is soon apparent":** Ritvo, 1990, p. 142.

206 **50 million black holes:** Mitchell Begelman, lecture at the American Museum of Natural History, Sept. 25, 2000.

206 **100 billion cells:** Sandra Blakeslee, "How Does the Brain Work?" *The New York Times,* Nov. 11, 2003.

BIBLIOGRAPHY

Adrian, E. D. "Science and Human Nature," *Supplement to Nature* (September 4, 1954), pp. 433–437.

Amacher, Peter. "The concepts of the pleasure principle and infantile erogenous zones shaped by Freud's neurological education," *Psychoanalytic Quarterly* (1974), pp. 218–223.

Armitage, Angus. *Sun, Stand Thou Still: The Life and Work of Copernicus the Astronomer.* New York: Henry Schuman, Inc., 1947.

Badash, Lawrence. "The Completeness of Nineteenth-Century Science," *Isis* (1972), pp. 48–58.

Bartusiak, Marcia. *Einstein's Unfinished Symphony: Listening to the Sounds of Space-Time.* Washington, D.C.: Joseph Henry Press, 2000.

Beecher, Willard. "The Myth of 'The Unconscious,'" *Individual Psychology Bulletin* (1950), pp. 99–110.

Bell, E.T. *Men of Mathematics: The Life and Achievements of the Great Mathematicians from Zeno to Poincaré.* New York: Touchstone, 1986 (reprinted from 1937).

———. *The Development of Mathematics.* New York: Dover Publications, Inc., 1992 (reprinted from 1945).

Benjamin, John D. "Approaches to a Dynamic Theory of a Development Round Table, 1949: 2. Methodological Considerations in the Validation and Elaboration of Psychoanalytical Personality Theory," *American Journal of Orthopsychiatry* (1950), pp. 139–156.

Bernfeld, Siegfried. "Freud's Earliest Theories and the School of Helmholtz," *Psychoanalytic Quarterly* (1944), pp. 341–62.

———. "Freud's Scientific Beginnings," *American Imago* (1949), pp. 163–196.

———. "Sigmund Freud, M.D., 1882–1885," *International Journal of Psycho-Analysis* (1951), pp. 204–217.

Bernhard, Carl Gustaf, Elisabeth Crawford, Per Sörbom, eds. *Science, Technology and Society in the Time of Alfred Nobel*. Oxford: Pergamon Press, 1982.

Bernstein, Jeremy. *Albert Einstein: And the Frontiers of Physics.* New York: Oxford University Press, 1996.

———, and Gerald Feinberg, eds. *Cosmological Constants: Papers in Modern Cosmology.* New York: Columbia University Press, 1986.

Blackmore, John T. *Ernst Mach: His Work, Life and Influence.* Berkeley: University of California Press, 1972.

Blum, Harold P. "From suggestion to insight, from hypnosis to psychoanalysis," in *Freud: Conflict and Culture*, edited by Michael S. Roth. New York: Alfred A. Knopf, 1998.

Bohr, Niels. *The Philosophical Writings of Niels Bohr, Volume I: Atomic Theory and the Description of Nature.* Woodbridge, Conn.: Ox Bow Press, 1987 (reprinted from 1934).

———. *The Philosophical Writings of Niels Bohr, Volume II: Essays 1932–1957 on Atomic Physics and Human Knowledge.* Woodbridge, Conn.: Ox Bow Press, 1987 (reprinted from 1958).

———. *The Philosophical Writings of Niels Bohr. Volume III: Essays 1958–1962 on Atomic Physics and Human Knowledge.* Woodbridge, Conn.: Ox Bow Press, 1987 (reprinted from 1963).

Boorstin, Daniel J. *The Discoverers.* New York: Vintage Books, 1985.

Bowers, Kenneth S., and Donald Meichenbaum, eds. *The Unconscious Reconsidered.* New York: John Wiley & Sons, 1984.

Brian, Dennis. *Einstein: A Life.* New York: John Wiley & Sons, Inc., 1996.

Brower, Daniel. "The Problem of Quantification in Psychological Science," *Psychological Review* (1949), pp. 325–333.

Bruck, Mark Anton. "The Concept of 'The Unconscious,'" *Individual Psychology Bulletin* (1950), pp. 81–98.

Bunge, Mario, and William R. Shea, eds. *Rutherford and Physics at the Turn of the Century.* New York: Dawson and Science History Publications, 1979.

Child, C. M., Kurt Koffka, John E. Anderson, John B. Watson, Edward Sapir, W. I. Thomas, Marion E. Kenworthy, F. L. Wells, William A. White. *The Unconscious: A Symposium.* New York: Alfred A. Knopf, 1927.

Clark, Ronald W. *Einstein: The Life and Times.* New York: Bard, 1999 (reprinted from 1972).

Clarke, Edwin, and L. S. Jacyna. *Nineteenth-Century Origins of Neuroscientific Concepts.* Berkeley: University of California Press, 1987.

Cohen, I.B. "Roemer and the First Determination of the Velocity of Light (1676)," *Isis* (April 1940), pp. 327–79.

Coles, Peter, ed. *The Routledge Critical Dictionary of the New Cosmology.* New York: Routledge, 1999.

Cranefield, Paul F. "The Organic Physics of 1847 and the Biophysics of Today," *Journal of the History of Medicine and Allied Sciences* (1957), pp. 407–23.

Crews, Frederick. *The Memory Wars: Freud's Legacy in Dispute.* New York: New York Review of Books, 1995.

Crews, Frederick C. *Unauthorized Freud: Doubters Confront a Legend.* New York: Viking, 1998.

Darwin, Charles. *On The Origin Of Species By Means Of Natural Selection, or, The Preservation Of Favoured Races In The Struggle For Life.* The Project Gutenberg Literary Archive Foundation: www.gutenberg.net, release 1228, March 1998.

Darwin, Francis, ed. *The Life and Letters of Charles Darwin.* New York: Basic Books, Inc., 1959.

Descartes, René. *The Philosophical Writings of Descartes, Volume I,* translated by John Cottingham, Robert Stoothoff, Dugald Murdoch. Cambridge: Cambridge University Press, 1985.

———. *Discourse On The Method Of Rightly Conducting One's Reason And Of Seeking Truth In The Sciences.* The Project Gutenberg Literary Archive Foundation: www.gutenberg.net, release 59, March 1993.

———. *Meditations on First Philosophy: with Selections from the Objections and Replies*, translated and edited by John Cottingham. Cambridge: Cambridge University Press, 1996.

Dictionary of Scientists, A. Oxford: Oxford University Press, 1999.

Dilman, Ilham. "The Unconscious," *Mind* (October 1959), pp. 446–473.

Dingle, Herbert. "Scientific and Philosophical Implications of the Special Theory of Relativity," in *Albert Einstein: Philosopher-Scientist*, ed. Paul Arthur Schilpp. New York: MJF Books, 2001 (reprinted from 1949).

Eddington, A.S. *Space Time and Gravitation*. Cambridge: Cambridge University Press, 1921 (reprinted from 1920).

Edelson, Marshall. *Psychoanalysis: A Theory in Crisis.* Chicago: University of Chicago Press, 1990.

Eidelberg, Ludwig. "The concept of the unconscious," *Psychiatric Quarterly* (1953) pp. 563–587.

Einstein, A. "The foundation of the general theory of relativity" (1916), in *The Principle of Relativity: A Collection of Original Memoirs on the Special and General Theory of Relativity*, by H.A. Lorentz et al. New York: Dover Publications, Inc., 1952 (reprinted from 1923).

———. "Cosmological considerations on the general theory of relativity" (1917), in *The Principle of Relativity: A Collection of Original Memoirs on the Special and General Theory of Relativity*, by H.A. Lorentz et al. New York: Dover Publications, Inc., 1952 (reprinted from 1923).

———. "Do gravitational fields play an essential part in the structure of the elementary particles of matter?" (1919), in *The Principle of Relativity: A Collection of Original Memoirs on the Special and General Theory of Relativity*, by H.A. Lorentz et al. New York: Dover Publications, Inc., 1952 (reprinted from 1923).

Einstein, Albert. "Autobiographical Notes" (1949), in *Albert Einstein:*

Philosopher-Scientist, edited by Paul Arthur Schilpp. New York: MJF Books, 2001 (reprinted from 1949).

———. *Albert Einstein, The Human Side: New Glimpses from His Archives*, selected and edited by Helen Dukas and Banesh Hoffmann. Princeton: Princeton University Press, 1979.

———. *The Collected Papers of Albert Einstein, Volume 1, The Early Years: 1879–1902*, translated by Anna Beck, in consultation with Peter Havas. Princeton: Princeton University Press, 1987.

———. *Ideas and Opinions*. New York: Bonanza Books, 1988 (reprinted from 1954).

———. *Relativity: The Special and the General Theory*, translated by Robert W. Lawson. New York: Three Rivers Press, 1995 (reprinted from 1920).

Ekeland, Ivar. *Mathematics and the Unexpected*. Chicago: The University of Chicago Press, 1988.

———. *The Broken Dice, and Other Mathematical Tales of Chance*, translated by Carol Volk. Chicago: The University of Chicago Press, 1993.

Ellenberger, Henri. "The unconscious before Freud," *The Bulletin of the Menninger Clinic* (1957), pp. 3–15.

Ellenberger, Henri F. *The Discovery of the Unconscious: The History and Evolution of Dynamic Psychiatry*. New York: Basic Books, Inc., 1970.

Erikson, Erik H. "The First Psychoanalyst," *The Yale Review* (September 1956), pp. 40–62.

Feigl, Herbert, and Michael Scriven. *Minnesota Studies in the Philosophy of Science, Volume I: The Foundations of Science and the Concepts of Psychology and Psychoanalysis*. Minneapolis: University of Minnesota Press, 1956.

Field, G. C., F. Aveling, and John Laird. "Is the Conception of the Unconscious of Value in Psychology?," *Mind* (1922), pp. 413–442.

Fölsing, Albrecht. *Albert Einstein*. New York: Penguin Books, 1998.

Frank, Philipp. *Modern Science and Its Philosophy*. Cambridge, Mass.: Harvard University Press, 1949.

———, ed. *The Validation of Scientific Theories*. Boston: The Beacon Press, 1956.

———. *Philosophy of Science: The Link Between Philosophy and Science.* Englewood Cliffs, N.J.: Prentice-Hall, Inc., 1957.

———. *Einstein: His Life and Times*, translated by George Rosen, edited and revised by Shuichi Kusaka. New York: Da Capo, 1989 (reprinted from 1953).

French, A. P. *Einstein: A Centenary Volume.* Cambridge, Mass.: Harvard University Press, 1979.

Frenkel-Brunswick, Else. "Psychoanalysis and the Unity of Science," *Proceedings of the American Academy of Arts and Sciences* (March 1954), pp. 273–347.

Freud, Sigmund. *The Standard Edition of the Complete Psychological Works of Sigmund Freud*, translated under the general editorship of James Strachey, in collaboration with Anna Freud, assisted by Alix Strachey and Alan Tyson. London: The Hogarth Press and the Institute of Psycho-Analysis, 1953–1966.

———. *The Origins of Psycho-Analysis: Letters to Wilhelm Fliess, Drafts and Notes: 1887–1902*, edited by Maria Bonaparte, Anna Freud, Ernst Kris, authorized translation by Eric Mosbacher and James Strachey, introduction by Ernst Kris. New York: Basic Books, Inc., 1954.

———. *Collected Papers*, translation under the supervision of Joan Riviere. New York: Basic Books, Inc., 1959.

———. *Letters of Sigmund Freud*, selected and edited by Ernst L. Freud, translated by Tania and James Stern. New York: Basic Books, Inc., 1960.

———, and Sándor Ferenczi. *The Correspondence of Sigmund Freud and Sándor Ferenczi: Volume 2, 1914–1919*, edited by Ernst Falzeder and Eva Brabant, with the collaboration of Patrizia Giampieri-Deutsch, translated by Peter T. Hoffer. Cambridge, Mass.: Harvard University Press, 1996.

Frey-Rohn, Liliane. *From Freud to Jung: A Comparative Study of the Psychology of the Unconscious*, translated by Fred E. Engreen and Evelyn E. Engreen. New York: G. P. Putnam's Sons, 1974.

Friedman, Herbert. *The Astronomer's Universe.* New York: Ballantine Books, 1990.

Friedmann, Aleksandr. "On the Curvature of Space" (1924), translated by Brian Doyle, in *Source Book in Astronomy and Astrophysics 1900–1975*, edited by Kenneth R. Lang and Owen Gingerich. Cambridge, Mass.: Harvard University Press, 1979.

Gabbard, Glen O. "Mind and brain in psychiatric treatment," *The Bulletin of the Menninger Clinic* (1994), pp. 427–446.

Gabriel, Yiannis. "The Fate of the Unconscious in the Human Sciences," *Psychoanalytic Quarterly*, v. 51, 1982, April, p. 246.

Galilei, Galileo. *Discoveries and Opinions of Galileo*, translated with an introduction and notes by Stillman Drake. New York: Anchor Books, 1957.

———. *Sidereus Nuncius, or The Sidereal Messenger*, translated and with introduction, conclusion, and notes by Albert Van Helden. Chicago: University of Chicago Press, 1989.

Galison, Peter. "Einstein's Clocks: The Place of Time," in *The Best American Science Writing 2000,* edited by James Gleick and Jesse Cohen. New York: Harper Collins, 2000.

Gay, Peter. *Freud: A Life for Our Time.* New York: W. W. Norton & Company, 1988.

Giacconi, Riccardo, Herbert Gursky, Frank R. Paolini, and Bruno B. Rossi. "Evidence for X-Rays from Sources outside the Solar System" (1962), in *Source Book in Astronomy and Astrophysics 1900–1975*, edited by Kenneth R. Lang and Owen Gingerich. Cambridge, Mass.: Harvard University Press, 1979.

Gill, Merton. "The Present State of Psychoanalytic Theory," *Journal of Abnormal and Social Psychology* (1959), pp. 1–8.

Giora, Zev. *The Unconscious and the Theory of Psychoneuroses.* New York: New York University Press, 1989.

Glasser, Otto. *Dr. W. C. Röntgen.* Springfield, Ill.: Charles C. Thomas, 1958 (reprinted from 1945).

Gray, A. "Lord Kelvin," *Nature* (June 18, 1896), pp. 151–52.

———. "Lord Kelvin's Jubilee," *Nature* (June 25, 1896), pp. 173–81.

Gray, Andrew. *Lord Kelvin: An Account of His Scientific Life and Work.* New York: E. P. Dutton & Co., 1908.

Grünbaum, Adolf. *The Foundations of Psychoanalysis: A Philosophical Critique.* Berkeley: University of California Press, 1985.

———. "A century of psychoanalysis: critical retrospect and prospect," in *Freud: Conflict and Culture*, edited by Michael S. Roth. New York: Alfred A. Knopf, 1998.

Hawking, Stephen, and Roger Penrose. *The Nature of Space and Time.* Princeton: Princeton University Press, 1996.

Hawking, Stephen and Werner Israel, eds. *300 Years of Gravitation.* Cambridge: Cambridge University Press, 1987.

Hawkins, Michael. *Hunting Down the Universe.* Reading, Mass.: Addison-Wesley, 1997.

Heilbron, J.L. "Fin-de-Siècle Physics," in *Science, Technology and Society in the Time of Alfred Nobel*, edited by Carl Gustaf Bernhard, Elisabeth Crawford, Per Sörbom. Oxford: Pergamon Press, 1982.

Heisenberg, Werner. *The Physical Principles of the Quantum Theory*, translated by Carl Eckart and F.C. Hoyt. New York: Dover Publications, Inc., 1949 (reprinted from 1930).

———. *Physics and Philosophy.* New York: Harper & Brothers, 1958.

———. *Physics and Beyond: Encounters and Conversations.* New York: Harper & Row, 1971.

———. *Across the Frontiers*, translated by Peter Heath. New York: Harper & Row, 1974.

Herzog, Patricia S. *Consciousness and Unconsciousness: Freud's Dynamic Distinction Reconsidered.* Madison, Conn.: International Universities Press, Inc., 1991.

Hilgard, Ernest R., Lawrence S. Kubie, E. Pumpian-Mindlin. *Psychoanalysis as Science: The Hixon Lectures on the Scientific Status of Psychoanalysis*, edited by E. Pumpian-Mindlin. Stanford: Stanford University Press, 1952.

Hoffman, Banesh. *The Strange Story of the Quantum.* New York: Dover Publications, Inc., 1959 (reprinted from 1947).

Holt, Robert R. "Freud's Mechanistic and Humanistic Images of Man," *Psychoanalysis and Contemporary Science* (1972), pp. 3–24.

Holton, Gerald. "Introduction: Einstein and the Shaping of Our Imagination" (1982), in *Albert Einstein: Historical and Cultural Perspectives*, edited by Gerald Holton and Yehuda Elkana. Mineola, N.Y.: Dover, 1997 (reprinted from 1982).

———. *Thematic Origins of Scientific Thought: Kepler to Einstein.* Cambridge, Mass.: Harvard University Press, 1988.

———. *Science and Anti-Science.* Cambridge, Mass.: Harvard University Press, 1993.

———. *Einstein, History, and Other Passions.* Woodbury, N.Y.: American Institute of Physics, 1995.

———. *The Advancement of Science, and Its Burdens.* Cambridge, Mass.: Harvard University Press, 1998a.

———. *The Scientific Imagination.* Cambridge, Mass.: Harvard University Press, 1998b.

———, and Yehuda Elkana, eds. *Albert Einstein: Historical and Cultural Perspectives.* Mineola, N.Y.: Dover, 1997 (reprinted from 1982).

Hook, Sidney, ed. *Psychoanalysis, Scientific Method and Philosophy.* New York: New York University Press, 1959.

Hoskin, Michael. *Stellar Astronomy.* Bucks, England: Science History Publications, 1982.

Hubble, Edwin. "A Relation between Distance and Radial Velocity among Extra-Galactic Nebulae," *Proceedings of the National Academy of Sciences* (March 15, 1929), pp. 168–173.

———. "Cepheids in Spiral Nebulae," *Observatory* (May 1925), pp. 139–142 (reprinted from *Publications of the American Astronomical Society* [1925], pp. 261–264).

James, William. *The Principles of Psychology.* New York: Dover, 1950 (reprinted from 1890).

Jones, Ernest. "Why is the 'Unconscious' Unconscious?—III," *British Journal of Psychology* (October 1918), pp. 247–256.

———. *The Life and Work of Sigmund Freud.* New York: Basic Books, Inc., 1953–1957.

Kelly, William L. *Psychology of the Unconscious: Mesmer, Janet, Freud, June, and Current Issues.* Buffalo, N.Y.: Prometheus Books, 1991.

Kelvin, Lord. *Baltimore Lectures on Molecular Dynamics and the Wave Theory of Light*. London: C.J. Clay and Sons, 1904.

Kolb, Rocky. *Blind Watchers of the Sky: The People and Ideas That Shaped Our View of the Universe.* New York: Addison-Wesley, 1996.

Kubie, Lawrence S. "The Fallacious Use of Quantitative Concepts in Dynamic Psychology," *Psychoanalytic Quarterly* (1947), pp. 507–518.

Lang, Kenneth R., and Owen Gingerich, eds. *Source Book in Astronomy and Astrophysics 1900–1975*. Cambridge, Mass.: Harvard University Press, 1979.

Learner, Richard. *Astronomy Through the Telescope.* New York: Van Nostrand Reinhold Company, 1981.

Lewin, Kurt. "The Conflict Between Aristotelian and Galileian Modes of Thought in Contemporary Psychology," *Journal of General Psychology* (1931), pp. 141–177.

Livingston, Dorothy Michelson. *The Master of Light: A Biography of Albert A. Michelson*. New York: Charles Scribner's Sons, 1973.

London, Ivan D. "Psychologists' Misuse of the Auxiliary Concepts of Physics and Mathematics," *Psychological Review* (1944), pp. 266–291.

Lorentz, H. A. "Michelson's Interference Experiment" (1895), in *The Principle of Relativity: A Collection of Original Memoirs on the Special and General Theory of Relativity,* by H.A. Lorentz et al. New York: Dover Publications, Inc., 1952 (reprinted from 1923).

———, A. Einstein, H. Minkowski and H. Weyl. *The Principle of Relativity: A Collection of Original Memoirs on the Special and General Theory of Relativity*, translated by W. Perrett and G. B. Jeffery. New York: Dover Publications, Inc., 1952 (reprinted from 1923).

Lubbock, Constance A., ed. *The Herschel Chronicle: The Life-Story of William Herschel and His Sister Caroline Herschel*. London: Cambridge University Press, 1933.

Mach, Ernst. *The Science of Mechanics: A Critical and Historical Account of Its Development*, translated by Thomas J. McCormack. La Salle, Ill.: The Open Court Publishing Company, 1974 (reprinted from 1893, 1902, 1919, 1942).

Margetts, Edward L. "The concept of the unconscious in the history of medical psychology," *Psychiatric Quarterly* (1953), pp. 115–138.

Masson, Jeffrey Moussaieff, translator and editor. *The Complete Letters of Sigmund Freud to Wilhelm Fliess 1887–1894*. Cambridge, Mass.: The Belknap Press, 1985.

Max, D. T. "Two Cheers for Darwin," *American Scholar* (Spring 2003), pp. 63–74.

Michelson, A. A. *Studies in Optics.* Chicago: The University of Chicago Press, 1927.

Miller, Arthur I. *Imagery in Scientific Thought: Creating Twentieth-Century Physics.* Boston: Birkhäuser, 1984.

———. *Frontiers of Physics: 1900–1911*. Boston: Birkhäuser, 1986.

———. *Albert Einstein's Special Theory of Relativity: Emergence (1905) and Early Interpretation (1905–1911)*. New York: Springer-Verlag, 1997.

Mujeeb-ur-Rahman, Md., ed. *The Freudian Paradigm: Psychoanalysis and Scientific Thought*. Chicago: Nelson-Hall, 1977.

Nelson, Benjamin, ed. *Freud and the Twentieth Century.* Cleveland: Meridian Books, 1957.

Neu, Jerome, ed. *The Cambridge Companion to Freud*. Cambridge, England: Cambridge University Press, 1991.

Newton, Isaac. *The Principia: Mathematical Principles of Natural Philosophy*, translated by I. Bernard Cohen and Anne Whitman. Berkeley: University of California Press, 1999.

Newton, Sir Isaac. *Mathematical Principles of Natural Philosophy*, in *Great Books of the Western World*, editor in chief Mortimer J. Adler, v. 32. Chicago: Encyclopaedia Britannica, Inc.: 1952.

Nicoll, Maurice. "Why is the 'Unconscious' Unconscious?—I," *British Journal of Psychology* (October 1918), pp. 230–235.

Nitske, W. Robert. *The Life of Wilhelm Conrad Röntgen: Discoverer of the X Ray.* Tucson: The University of Arizona Press, 1971.

North, John. *The Norton History of Astronomy and Cosmology.* New York: W. W. Norton & Company, 1994.

Opatow, Barry. "The Real Unconscious: Psychoanalysis as a Theory of

Consciousness," *Journal of the American Psychoanalytic Association* (1997), pp. 865–890.

Oppenheimer, J. R., and H. Snyder. "On Continued Gravitational Contraction," *Physical Review* (September 1, 1939), pp. 455–459.

Overbye, Dennis. *Einstein in Love: A Scientific Romance.* New York: Viking, 2000.

Pais, Abraham. '*Subtle is the Lord...*': *The Science and the Life of Albert Einstein.* Oxford: Oxford University Press, 1982.

———. *Inward Bound: Of Matter and Forces in the Physical World.* Oxford: Oxford University Press, 1988.

Phillips, Adam. *Darwin's Worms: On Life Stories and Death Stories.* New York: Basic Books, 2001.

Planck, Max. *Where Is Science Going?* Woodbridge, Conn.: Ox Bow Press, 1981 (reprinted from 1933)

Poincaré, H. *Science and Hypothesis.* New York: Dover Publications, Inc., 1952 (reprinted from 1905).

———. *The Foundations of Science: Science and Hypothesis; The Value of Science; Science and Method*, translated by George Bruce Halsted. Washington: University Press of America, 1982 (reprinted from 1913, 1946).

Popper, Karl. *The Logic of Scientific Discovery.* New York: Harper & Row, 1968.

———. *Unended Quest.* London: Routledge Classics, 2002 (reprinted from 1974 and 1992).

Rapaport, David, and Merton M. Gill. "The Points of View and Assumptions of Metapsychology," *The International Journal of Psycho-Analysis* (1959), pp. 153–162.

Reiser, Morton F. "Converging Sectors of Psychoanalysis and Neurobiology: Mutual Challenge and Opportunity," *Journal of the American Psychoanalytic Association* (1985), pp. 11–34.

Ritvo, Lucille B. "Darwin as the source of Freud's neo-Lamarckianism," *Journal of the American Psychoanalytic Association* (1965), pp. 499–517.

———. "The impact of Darwin on Freud," *Psychoanalytic Quarterly* (1974), pp. 177–192.

———. *Darwin's Influence on Freud: A Tale of Two Sciences.* New Haven: Yale University Press, 1990.

Rivers, W. H. R. "Why is the 'Unconscious' Unconscious?—II," *British Journal of Psychology* (October 1918), pp. 236–246.

Roseveare, N. T. *Mercury's Perihelion from Le Verrier to Einstein.* Oxford: Clarendon Press, 1982.

Roth, Michael S, ed. *Freud: Conflict and Culture.* New York: Alfred A. Knopf, 1998.

Sacks, Oliver. "The Other Road: Freud as Neurologist," in *Freud: Conflict and Culture*, edited by Michael S. Roth. New York: Alfred A. Knopf, 1998.

Schilpp, Paul Arthur, ed. *Albert Einstein: Philosopher-Scientist.* New York: MJF Books, 2001 (reprinted from 1949).

Schore, Allan N. "A Century After Freud's Project: Is a Rapprochement Between Psychoanalysis and Neurobiology at Hand?," *Journal of the American Psychoanalytic Association* (1997), pp. 807–840.

Schwarzschild, Karl. "On the Gravitational Field of a Point Mass according to the Einsteinian Theory" (1916), translated by Brian Doyle, in *Source Book in Astronomy and Astrophysics 1900–1975*, edited by Kenneth R. Lang and Owen Gingerich. Cambridge, Mass.: Harvard University Press, 1979.

Scriven, Michael. "Explanation and Prediction in Evolutionary Theory," *Science* (August 1959), pp. 477–482.

Serota, Herman M. "The ego and the unconscious: 1784–1884," *Psychoanalytic Quarterly* (1974), pp. 224–242.

Shope, Robert K. "Physical and Psychic Energy," *Philosophy of Science* (March 1971), pp. 1–12.

Shuey, Herbert. "Recent Trends in Science and the Development of Modern Typology," *The Psychological Review* (May 1934), pp. 207–235.

Singer, Charles. *A Short History of Science to the Nineteenth Century.* Mineola, N.Y.: Dover Publications, Inc., 1997 (reprinted from 1943).

Solms, Mark. "New Findings on the Neurological Organization of

Dreaming: Implications for Psychoanalysis," *Psychoanalytic Quarterly* (1995), pp. 43–67.

Spangenburg, Ray, and Diane K. Moser. *On the Shoulders of Giants: The History of Science in the Eighteenth Century.* New York: Facts on File, 1993.

Stachel, John. "Einstein and ether drift experiments," *Physics Today* (May 1987), pp. 45–47.

———, ed. *Einstein's Miraculous Year: Five Papers That Changed the Face of Physics.* Princeton: Princeton University Press, 1998.

Sterba, Richard F. "The humanistic wellspring of psychoanalysis," *Psychoanalystic Quarterly* (1974), pp. 167–176.

Sulloway, Frank J. *Freud, Biologist of the Mind.* New York: Basic Books, Inc., 1979.

Swenson, Lloyd S., Jr. "Michelson and measurement," *Physics Today* (May 1987), pp. 24–30.

Thompson, Silvanus P. *The Life of William Thomson, Baron Kelvin of Largs.* London: Macmillan and Co., Limited, 1910.

Turner, Michael S., and J. Anthony Tyson. "Cosmology at the Millenium," *Reviews of Modern Physics* (1999), pp. 145–164.

Van Helden, Albert. "The Telescope in the Seventeenth Century," *Isis* (1974), pp. 38–58.

———. "Conclusion," in *Sidereus Nuncius, or The Sidereal Messenger,* translated and with introduction, conclusion, and notes by Albert Van Helden. Chicago: University of Chicago Press, 1989.

Waelder, Robert. *Basic Theory of Psychoanalysis.* New York: International Universities Press, Inc., 1960.

Watson, John B. "The Unconscious of the Behaviorist," in *The Unconscious: A Symposium,* by C.M. Child, et al. New York: Alfred A. Knopf, 1927.

Weaver, Warren. "Science and People," *Science* (December 30, 1955), pp. 1255–1259.

———. "Science and the Citizen," *Science* (December 13, 1957), pp. 1225–1229.

Weinberg, Steven. "Newtonianism and today's physics," in *Three Hun-*

dred Years of Gravitation, edited by Stephen Hawking and Werner Israel. Cambridge: Cambridge University Press, 1987.

Weiner, Charles. "Who Said It First?," *Physics Today* (August 1968), p. 9.

Westfall, Richard S. *Never at Rest: A Biography of Isaac Newton*. Cambridge: Cambridge University Press, 1980.

Wheeler, John Archibald. "Our Universe: The Known and the Unknown," *American Scientist* (Spring 1968), pp. 1–20.

Whyte, L. L. *The Unconscious Before Freud*. New York: St. Martin's Press, 1978 (reprinted from 1960).

Will, Clifford M. *Was Einstein Right?: Putting General Relativity to the Test*. New York: Basic Books, 1993.

Williams, Henry Smith. *A History of Science*. The Project Gutenberg Literary Archive Foundation: www.gutenberg.net, releases 1705–1708, April 1999 (reprinted from 1904–1910).

Wollheim, Richard, ed. *Freud: A Collection of Critical Essays*. Garden City, N.Y.: Anchor Books, 1974.

———. *Sigmund Freud*. New York: Cambridge University Press, 1990.

Wolman, Benjamin B. *Logic of Science in Psychoanalysis*. New York: Columbia University Press, 1984.

Woolf, Harry. *Some Strangeness in the Proportion: A Centennial Symposium to Celebrate the Achievements of Albert Einstein*. Reading, Mass.: Addison-Wesley Publishing Company, Inc., 1980.

Zajonc, Arthur. *Catching the Light: The Entwined History of Light and Mind*. New York: Oxford University Press, 1995.

INDEX

Académie des Sciences, 136
Age of Discovery, 2–3, 64
Almagest (Ptolemy), 60
Ames, J. S., 176–77
animalcules, 2, 62, 159
animal magnetism, 47
animal spirits, 121, 122, 126, 127, 190
Annalen der Physik, 69, 82
Anna O., 51, 52, 75, 138
anti-Semitism, 35–36, 53
Archives de neurologie, 141
Aristotle, 24, 31, 34, 119, 120, 125, 154, 165, 190
  motion and, 18, 20–22, 28
  on terrestrial vs. celestial matter, 44–45
Astronomia nova (*New Astronomy*) (Kepler), 91–92
atoms, 157, 174
Augustine, Saint, 73
August P., 115–18, 142
"Autobiographical Notes" (Einstein), 153–54, 171

Baraduc, Dr., 58
Becquerel, Henri, 157
Bell, Charles, 38
Bell Telephone Laboratories, 193, 194
Berliner Physikalische Gesellschaft, 126, 127, 165
Bernard of Chartres, 61, 63
Bernheim, Hippolyte, 137–40
Besso, Michele, 23, 28, 32, 83, 166, 171, 173
big bang theory, 192, 193–94, 196, 206
black holes, 197–98, 205, 206
Brahe, Tycho, 89–90, 91, 173
brain, 36–44, 49–50, 54, 58, 123, 134, 140, 154, 180
  fibers in, 36–37, 39–41, 48, 73

brain (*cont.*)
"globular" hypothesis of, 38–39
mind vs., 44, 49–50, 118, 123
Breuer, Josef, 50–51, 52, 75–76, 138, 163, 201
Freud and, 75, 76, 142–43, 182
Brücke, Ernst, 126–27, 133, 134, 142, 198

Cäcilie M., 52
California Institute of Technology, 199–200
Carus, Karl Gustav, 74
catharsis, 51, 75, 200
celestial vault, 19, 25
cell theory, 39–40
Chandrasekhar, Subrahmanyan, 197
Chandra X-ray telescope, 205
Chaplin, Charlie, 188
Charcot, Jean-Martin, 116, 128–38, 140–41
hysteria research of, 128–33, 136–37, 141
children, sexuality of, 183
Clark University, 183
Claus, Carl, 125, 203
clocks, electrical, coordination of, 70–72
cocaine, 134, 146
compasses, 9–10, 34, 85
Comte, Auguste, 164, 166
consciousness, 127, 206
Descartes on, 5, 73, 123, 160–63
Freud on, 43, 49, 72–73, 154, 167
Contemporary Review, 179–80
Copernicus, Nicolaus, 18–21, 28, 88, 89, 92, 159, 189
"Cosmological Considerations" (Einstein), 111–12
cosmological constant, 112–14
cosmology, 6, 114, 192–98
psychoanalysis vs., 202–3
counter-transference, 201
Curie, Marie, 97, 157–58
Cygnus X-I, 197–98

Dampier, William, 64
Darwin, Charles, 124–26, 147, 148, 158, 162, 182, 189, 200, 203
De humani corporis fabrica (Vesalius), 120–21
De la suggestion et de ses applications à la thérapeutique (Bernheim), 137–38
De magnete, magnetisque corporibus, et de magno magnete tellure (Gilbert), 91–92
De motu cordis et sanguinis in animalibus (Harvey), 122
De revolutionibus orbium coelestium (Copernicus), 19, 88
Descartes, René, 12, 24, 43, 45–46, 60, 63–64, 74, 121–23, 159–64, 205
consciousness and, 6, 73, 123, 160–63
Descent of Man, The (Darwin), 124, 125
Dialogue Concerning the Two Chief World Systems (Galileo), 20–22, 46
Discourses and Mathematical Demonstrations Concerning Two New Sciences (Galileo), 161
dissections, 120–22, 136, 160
dreams, 53, 147, 148, 176, 181, 182, 184, 201
Drei Abhandlungen zur Sexualtheorie (Freud), 183
Du Bois-Reymond, Emil, 126, 142, 164–65, 166, 198

dynamo, first, 27, 171
Dyson, Frank, 109

Earth, 14–16, 24, 45, 92–93, 195, 205
as center of universe, 18–20, 189
motion of, 18–21, 45, 69, 113
Eddington, Arthur, 106, 176, 179–80, 184, 189, 197
Edison, Thomas, 57–58
Einstein, Albert, 9–11, 22–34, 57, 67–72, 81–88, 93–114, 153–56, 165–79, 184–92, 199, 204–7
Besso's friendship with, 23, 28, 32, 83, 166, 171, 173
cosmological constant and, 112–14
education of, 10–11, 98, 99
ether research and, 10, 11, 17–18, 22–23, 30–31, 67, 68, 77
falling-man image of, 81–82, 84, 95–96
father of, 9, 10, 27, 85
Freud's correspondence with, 187–88
Freud's meeting with, 1, 185
Freud vs., 1–2, 5–6, 77–78
Hubble visited by, 114
Mach's influence on, 28–30, 68, 69–70, 163, 169–71, 174
marriage of, 23, 96
Oxford lecture of, 170, 173–74
patent office job of, 11, 23, 72, 83–84, 96, 98, 167
Poincaré's influence on, 69–72
see also relativity theory; relativity theory, general; relativity theory, special
electricity, 10, 11, 27, 58, 171–72, 174
electrodynamics, 30, 69, 171
"Electromagnetic Phenomena in a System Moving with Any Velocity Smaller Than That of Light" (Lorentz), 69
electromagnetic spectrum, 178, 193–95, 205
electromagnetism, 27, 28, 30, 34, 69, 172, 174
light and, 10, 11, 13, 18, 26–28, 69, 178, 197
Elektrotherapie (Erb), 135
Elisabeth von R., 139–40
Engels, Friedrich, 158
Eötvös, Baron Lóránt, 166
equivalence principle, 97
Erb, William, 135
ether, 13–18, 190
Einstein's research and, 10, 11, 17–18, 22–23, 30–31, 67, 68, 77
Michelson's experiments and, 14–17, 68
X-rays and, 57–58, 178
Euclid, 99
Exner, Sigmund, 48
expanding universe, 112, 113, 176, 192, 196
Exposition du système du monde (*The System of the World*) (Laplace), 94, 196

Faraday, Michael, 27, 157, 171–74
fibers, in brain, 36–37, 39–41, 48, 73
FitzGerald, George Francis, 16
Fliess, Wilhelm, 42–44, 53, 57, 147, 148, 167, 177, 201
forces, 158–62, 172
"Foundation of the General Theory of Relativity, The" (Einstein), 107
Freud, Martha Bernays, 129–33, 201–2

Freud, Sigmund, 35–37, 39–44, 47–54, 67, 72–78, 115–49, 154–56, 163, 165–68, 180–91, 198–207
Breuer's collaboration with, 75, 76, 142–43, 182
consciousness as viewed by, 43, 49, 72–73, 154, 167
dream of, 53, 147
Einstein compared with, 1–2, 5–6, 77–78
Einstein's correspondence with, 187–88
Einstein's meeting with, 1, 185
father of, 35–36, 53, 147
Fliess's correspondence with, 42–44, 53, 57, 147, 148, 167, 177
homesickness of, 129, 131
hypnosis used by, 138–40, 145
in Paris, 116–18, 127–37, 140
private practice of, 37, 41–42, 50, 134–36, 138–40, 143–46
professional crisis of, 129–30
on scientific creativity, 175–76
self-analysis of, 146–47
at University of Vienna, 40, 119, 125–27, 137, 148, 165–66, 183, 203
at Vienna General Hospital, 48, 116, 127, 133–34, 135
Friedmann, Aleksandr Aleksandrovich, 112
"Further Remarks on the Neuro-Psychoses of Defense" (Freud), 148–49

galaxies, 192, 205
Galen, 60, 73
Galileo Galilei, 2–3, 6, 18–22, 25, 28, 31–32, 60–64, 67, 95, 96, 161
heliocentric theory and, 18–21, 34, 46, 92
Jupiter's moons and, 2, 45, 159
new physics attempted by, 20–22, 46
relativity principle of, 22, 69
geometry, 46, 85–89, 92, 99–100, 105
Geometry (Descartes), 46
Gerlach, Joseph von, 39–40
Gesellschaft für positivistische Philosophie, 166
Gilbert, William, 91–92
"globular" hypothesis, 38–39
Goethe, Johann Wolfgang von, 123–24
Golgi, Camillo, 40
gravity, 47, 82, 84, 93–104, 190, 195, 206
cosmological constant and, 112
Newton's theory of, 5, 93–96, 103–4, 111, 160–61, 170, 193, 196, 197–98
relativity theory and
"*Grenzen des Naturerkennens, Die*" (Du Bois-Reymond), 165
Griesinger, Wilhelm, 74
Grossmann, Marcel, 98–99

Halley, Edmund, 93
Hartmann, Eduard von, 74
Harvey, William, 122
Hawking, Stephen, 192
heart, 121–22
Heisenberg, Werner, 174–75, 203–4
heliocentric theory, 18–21, 34, 46, 88, 89, 92
Helmholtz, Hermann von, 126
Herbart, Johann Friedrich, 47
Herder, Johann Gottfried von, 124

"Heredity and the Aetiology of the Neuroses" (Freud), 148–49
Herschel, William, 25, 28, 178
Hertz, Heinrich, 166, 178
histology, 38–39, 43
Hodgkin, Thomas, 38
Hooke, Robert, 71
Hubble, Edwin, 112–14, 192
human, 120–24
  animal vs., 123, 124, 160
  central nervous system of, 127, 134
  dissection of, 120–21
Huygens, Christian, 63
hypnoid states, 76
hypnosis, 47, 136–40
  Breuer's use of, 50–51, 138
  therapeutic potential of, 138–39, 145–46
hysteria, 41, 50–53, 136–46, 201
  Charcot's research into, 128–33, 136–37, 141
  grande vs. *petite,* 116–17
  hypnoid vs. defense, 76
  hypnosis in treatment of, 50–51
  symptoms of, 52, 53, 117, 131, 132, 136–37, 140–41

inner universe, 46–50
interferometers, 14–16, 66, 69
Interpretation of Dreams, The (*Die Traumdeutung*) (Freud), 148, 182, 198
inverse-square law, 92–93, 102
invisibility, second scientific revolution and, 5–6
island universes, 113
"Is the Conception of the Unconscious of Value in Psychology?" (symposium), 190–91

Jahrbuch der Radioaktivität und Elektronik, 82, 83, 84
James, William, 74
Jenseits des Lustprinzips (Freud), 202
Jews, 86
  anti-Semitism and, 35–36, 53
Joint Session of the Mind Association and the Aristotelian Society, 190–91
Jokes and Their Relationship to the Unconscious (Freud), 182
Jupiter, 95, 103
  moons of, 2, 24, 45, 71, 159

Kelvin, Lord (William Thomson), 11–16, 29, 65–66
Kepler, Johannes, 88–94, 101, 110, 111, 113, 173

Landau, Lev, 197
language, 72, 74
Laplace, Pierre-Simon de, 94–95, 102, 126, 162, 196, 197, 203
Laue, Max von, 108
Leeuwenhoek, Antonius von, 2–3, 6, 37–38, 62–63, 67, 159
Leibniz, Gottfried Wilhelm, 73
Le Verrier, Urbain-Jean-Joseph, 101–3
light, 17–18, 30–34, 57, 58, 157
  bending of, 100–101, 106, 107, 109, 110, 112
  electromagnetic waves of, 10, 11, 13, 18, 26–28, 69, 178, 197
  of massive stars, 196–98
  speed of, 22–28, 30–34, 68–71, 171, 197, 206
  X-rays and, 178–79, 184, 195–98

Lister, Joseph Jackson, 38–39
Logic of Scientific Discovery, The (Popper), 189–90, 191
Lorentz, Hendrik Antoon, 16–17, 68–69, 70, 106, 157
Lorentz-Einstein theory, 69

Mach, Ernst, 163–66, 168–71, 173
  Einstein influenced by, 28–30, 68, 69–70, 163, 169–71, 174
magnetism, 9–10, 11, 27, 92, 157, 171–72, 174
Maric, Mileva, 18, 23
Mars, orbit of, 89–91, 173
Marx, Karl, 158
mathematics, 47, 84–113, 161, 172, 176, 197, 198
  Descartes's views on, 46, 92
  Einstein's views on, 84–88, 95–111
  ether and, 16–19, 23
  Kepler's assumptions about, 88–94, 101, 110, 111
  relativity theory and, 68–69, 186, 192
  of velocity of light, 27–28
matter, 170
  Descartes on, 12, 45, 46, 73, 92
  motion and, 12–13, 17–18, 43, 44
  terrestrial vs. celestial, 44–45, 92
Max Kassowitz Institute for Children's Diseases, 134
Maxwell, James Clerk, 27–28, 65, 69, 172, 173, 174, 178
Mead, Richard, 47
mechanics, 12–13, 28–30, 163, 165, 170–74
Mechanik in ihrer Entwicklung, Die (Mach), 28–29
memory, 41, 73, 51–52, 140, 142–45
  fantasy in, 200
Mendenhall, T. C., 64
Mercury, orbit of, 101–5, 107
Mesmer, Franz Anton, 47
"Mesure du temps, La" (Poincaré), 70–71
Meynert, Theodor, 136
Michelson, Albert A., 14–17, 65, 66, 68, 69
Michelson-Morley experiment, 15–16, 17
microscopes, 2–4, 37–40, 48, 62–63, 67, 155–56, 204, 207
Millikan, Robert A., 158, 194
mind, 44, 179–80
  brain vs., 44, 49–50, 118, 123
  Descartes's views on, 73, 123
  Freud's views on, 42–43, 49–54, 136, 142–48, 154
  Herbart's views on, 47
moon, 19, 62, 93, 95, 195–96
Morley, Edward W., 15–16, 17
motion, 91, 159–62, 170, 205–6
  of Earth, 18–21, 45, 69, 113
  Galileo's views on, 20–22
  matter and, 12–13, 17–18, 43, 44
  Newton's views on, 12, 28–30, 43, 94, 110, 159
  predictability of, 94, 110, 111
  regularity of, 94, 110
Mount Wilson Observatory, 112–14

Napoleon, Emperor of France, 162
natural law, 88, 94, 102
natural philosophy, 19, 37–38, 63, 67, 78, 154–55, 161, 164–66, 179, 184, 189
natural selection, 124, 125, 158, 162

Nature, 66
Neptune, 95, 110
nervous system, neurons, 36–43, 48, 54, 67, 77
Freud's views on, 36–37, 41–42, 43, 49, 118, 137, 140–42
hypnosis and, 137
nervous impulse and, 126
of primitive fishes vs. humans, 127
"neue Art von Strahlen, Eine" (Röntgen), 56
neuroanatomy, Freud's research in, 36–37, 41, 42, 47–49, 118–21, 127, 130, 132–33, 140–41
neuron doctrine, 41, 42, 48
Newcomb, Simon, 65, 103, 104
Newton, Isaac, 5, 43, 46–47, 61, 63, 92–96, 108, 110, 113–14, 159–64, 179, 185, 188, 189, 205
apples-and-gravity meditation of, 82, 92–93
Einstein's overthrow of, 108, 109
gravity theory of, 5, 93–96, 103–4, 111, 160–61, 170, 193, 196, 197–98
mechanics of, 12, 28–30, 159, 170–74
New York Times, 57, 58, 108, 156, 198
"Note on the Unconscious in Psychoanalysis, A" (Freud), 190

objectivity, 155
observers:
clock coordination and, 71
Einstein on, 26–27, 30–34, 69
Galileo on motion and, 20–22
perceptual changes of, 69
star collapse and, 197
"On the Electrodynamics of Moving Bodies" (Einstein), 24, 82–83
"On the Gravitational Field of a Point Mass According to the Einsteinian Theory" (Schwarzschild), 196
On the Origin of Species (Darwin), 124, 125, 162, 203
"On the Psychical Mechanism of Hysterical Phenomena (Preliminary Communication)" (Breuer and Freud), 75–76
On the Revolution of Celestial Orbs (Copernicus), 88
Oppenheimer, J. Robert, 197
optics, 30, 69
Optics (Newton), 179
Outline of Psychoanalysis, An (Freud), 198, 199

Paris, Freud in, 116–18, 127–37
Pearson, Charles Henry, 162
perception, 6, 69, 173, 175, 178–80, 184
Philosophiae naturalis principia mathematica (Newton), 93–96, 160, 161
Philosophie des Unbewussten (*Philosophy of the Unconscious*) (Hartmann), 74
photography, 179, 184
Pinel, Philippe, 128, 135
Planck, Max, 186
planets, 159, 205
orbits of, 89–95, 101–5, 107, 173
Plato, 85, 119, 154, 155–56, 162, 163, 165, 190
Poincaré, Henri, 16, 68–72, 74, 113, 165

Popper, Karl, 189–90, 191
positivism, 164–75
  logical, 189–90, 193
posthypnotic suggestion, 137–40, 145
pressure technique, 139–40, 145–46
Princeton University, 193, 194
Prussian Academy, 104–5
Psyche (Carus), 74
psychoanalysis, 1, 6, 167, 180–81, 184, 189–92, 198–207
  deficiencies of, 198–202
  Einstein's views on, 187
  falsifiability criteria and, 202–3
  Freud's choice of term, 148–49
psychology, Freud's switch to, 118
"Psychology for Neurologists" (Freud), 41–44, 48–50, 118, 142, 148, 167, 198
Psychopathology of Everyday Life, The (Freud), 182
Ptolemy of Alexandria, 60
Pythagorean theorem, 87, 99

radioactivity, 158
radio waves, 178, 193–95, 206
Ramón y Cajal, Santiago, 41, 48
relativity, Galileo's principle of, 22, 69
relativity theory, 6, 153, 163, 169, 174–80, 196
relativity theory, general, 95–114, 166–67, 170, 176–77, 185–86, 192
  cosmological applications of, 111–14
  Eddington on, 179–80, 185
  gravity and, 95–104, 106, 170
  Laue's views on, 108
  news media's response to, 108–9
  postwar effects of, 109–10
  tests of, 100–101, 106–9
relativity theory, special, 30–34, 68–69, 106, 170–71
  as *Invariantentheorie* (invariant theory), 186
repression, 53, 76–77, 140, 145, 146, 147, 177–78, 181
resistance, 143–46, 177–78
Riemann, Bernhard, 100
Rogers, Ingles, 58
Roman Catholic Church, 46
Rømer, Ole, 24, 71
Röntgen, Bertha, 55, 56, 156
Röntgen, Wilhelm Conrad, 55–58, 156–57, 168, 181
Royal Astronomical Society, 109
Royal Institution, 172
Royal Society, 109, 110, 176

Salpêtrière, 116, 128–37, 140, 148
Saturn, 63, 95
Schwann, Theodor, 39
Schwarzschild, Karl, 196
science, 59–67, 153–56
  processes leading to, 153–55
Science, 16, 58, 203
Science and Hypothesis (Poincaré), 16
Science of Mechanics, The (Mach), 163
Scientific American, 109
scientific method, 2, 55–56, 59, 65, 77, 78, 135, 155, 162–63, 172–73
Scientific Revolution, 2–6, 55–56, 64, 159, 164
seduction theory, 200
"Seeing the Invisible" (Millikan), 194
self-analysis, 146–47
sexual abuse, 200
sexuality, 147, 183
Shakespeare, William, 4
Smith, Adam, 158
Snyder, Hartland S., 197
space, 180

absolute, 18, 22, 23, 29, 31, 57, 68, 167
curvature of, 100, 104, 110, 112, 175, 177–78, 179
spinal cord, 39, 40, 127, 134
stars, 2, 62, 113, 196–98
in celestial vault, 19, 25
fixed, 22, 25
velocities of, 113
steady state theory, 192, 193
Stokes, George Gabriel, 13
Studies on Hysteria (Breuer and Freud), 75, 76, 142–43, 144, 182
sun, 62, 113
eclipse of, 100–101, 106–9, 112, 176
gravity of, 101, 109
in heliocentric theory, 18–21, 34, 46, 88, 89, 92

telescopes, 2–4, 25, 60–63, 67, 92, 112–13, 155–56, 204, 205, 207
Thomas Aquinas, Saint, 88
Thomson, Joseph John, 65, 157, 176
thought, 36, 49, 58, 73, 153–55, 160, 168, 173, 174, 181, 184
3-degree signal, discovery of, 193–95
time, 23–27, 59, 100, 177–78, 180
absolute, 23–24, 68, 72, 167, 174–75
clock coordination and, 70–72
speed of light and, 26–27, 32–34
Times (London), 108, 109, 110, 185
Titian (Tiziano Vecellio), 120
Traité de mécanique céleste (*Celestial Mechanics*) (Laplace), 94, 102, 162
transference, 201
trauma, 50–51, 117, 139–40
repression of, 53, 76–77, 140, 145, 146
Treatise on Man (Descartes), 46
two-body problem, 102

"Über die Untersuchung des Ätherzustandes im magnetischen Felde" (Einstein), 10
unconscious, 49, 72–78, 190–91, 204–5
Freud's views on, 6, 49, 52–53, 72–73, 75–76, 175, 176, 177, 180–84, 190
illusion of absolute time in, 72
see also psychoanalysis; repression
"Unconscious, The" (Freud), 167
United States Physical Society, 176–77
uranium, 157
Uranus, 95, 103

velocity, defined, 32
Venus, 19–20, 102–3
Vesalius, Andreas, 120–22, 127
Vienna, University of, Freud at, 40, 119, 125–27, 137, 148, 165–66, 183, 203
Vienna Circle, 189
Vienna General Hospital, Freud at, 48, 116, 127, 133–34, 135
Vienna Institute of Physiology, 166
Vienna Medical Club, 52
Vienna Psychiatric Society, 36–37, 48, 137
Vienna Royal Academy of Science, 125
Vienna Society of Physicians, 115–18, 201
vitalism, 126

Waldeyer, Wilhelm, 41, 48
Wednesday Psychological Society, 182–83
Wheatstone, Charles, 13
Wheeler, John Archibald, 197
"When the cause ceases, the effect ceases" (Breuer and Freud), 75
Why War? (Einstein and Freud), 188
Wilhelm II, Kaiser of Germany, 57
World, The (Descartes), 45–46
Wren, Christopher, 60

X-rays, 55–59, 67, 156–57, 178–79, 181, 184, 195–98, 205

Zeeman, Pieter, 157